An Introduction to the Philosophy of Physics

*To Bas van Fraassen
of Princeton University,
who inspired me to want to become
a philosopher of science*

An Introduction to the Philosophy of Physics

Locality, Fields, Energy, and Mass

MARC LANGE

Blackwell Publishing

BLACKWELL PUBLISHING
350 Main Street, Malden, MA 02148-5020, USA
9600 Garsington Road, Oxford OX4 2DQ, UK
550 Swanston Street, Carlton, Victoria 3053, Australia

First published 2002 by Blackwell Publishing Ltd

5   2006

Library of Congress Cataloging-in-Publication Data has been applied for.

ISBN-13: 978-0-631-22500-3 (hardback)
ISBN-10: 0-631-22500-5 (hardback)
ISBN-13: 978-0-631-22501-0 (paperback)
ISBN-10: 0-631-22501-3 (paperback)

A catalogue record for this title is available from the British Library.

Set in 11 on 13 pt Baskerville
by Graphicraft Ltd, Hong Kong

The publisher's policy is to use permanent paper from mills that operate a
sustainable forestry policy, and which has been manufactured from pulp
processed using acid-free and elementary chlorine-free practices. Furthermore,
the publisher ensures that the text paper and cover board used have met
acceptable environmental accreditation standards.

For further information on
Blackwell Publishing, visit our website:
www.blackwellpublishing.com

Contents

Preface

> *The real difficulty lies in the fact that physics is a kind of metaphysics; physics describes "reality." But we do not know what "reality" is; we know it only by means of the physical description.*
>
> Einstein in 1935 (Howard 1989: 224)

A principal aim of the philosophy of physics is to work out what the universe must (or may) be like considering the remarkable success that various theories in physics have had in predicting our observations. In other words, the philosophy of physics is concerned with *interpreting* physical theories – with figuring out what they tell us about reality. As Einstein remarked, we must work from our best physical theories back to what reality is.

It might seem to you that there is nothing especially *philosophical* about this task – that it is the job of *physics* to tell us what reality is like. Actually, this job belongs to *both* physics and philosophy; as we will see, there is no sharp line dividing physics from the philosophy of physics. Necessarily, there is a fair amount of physics in this book, mostly drawn from electromagnetism, relativity, and basic quantum mechanics. But the way in which these theories are treated in physics classes and textbooks is typically quite different from the way in which they are analyzed in the philosophy of physics.

I know this difference first-hand. When I was an undergraduate, I originally concentrated on physics. As I worked my way through the sequence of required courses, I found that they were primarily concerned with teaching me how various scientific theories and mathematical techniques could be used to solve various problems, such as predicting a body's path under various conditions. Though learning how to do these things was interesting, no burning desire to do so had led me to concentrate on physics. Rather, I had wanted to know what the universe is like in its most fundamental respects: what sorts of things it is made of and how they work. The physics courses I took had no time for the questions that I found myself asking as I learned more physics; there was barely enough time to review each week's set of homework problems. But more than time pressure was involved. When questions of the kind *I* thought important did arise, they were often belittled with a hostility that quite surprised me. I know now (though didn't know then) that not all physicists would have responded to my questions in this way. But I also know now (though didn't know then) that my experience was not unique; others who at some point in their educations moved from physics to philosophy underwent searing experiences very similar to mine.

Here is one of the questions I wondered about. We are all familiar with the fact that physics talks about *fields* of force. A force field is exerted by something, occupies a region of space that may be empty of matter, and is able to push things around. Force fields have even entered science fiction: in *Star Trek*, a cell in the ship's brig has no bars, but instead is surrounded by an invisible force field. Physics also refers to *lines of force* curving through space. For example, magnetic lines of force are revealed by iron filings (see figure P.1); compass needles line up with them. Likewise, physics speaks of *potentials*, as when a proton is characterized as trapped in an atom's nucleus by virtue of lying at the bottom of a "potential well" (figure P.2). Now the question I wanted answered was: are fields (or lines of force, or potentials) real things or not?

This question concerns how we should *interpret* a remark describing the electric field in some region. Take the remark "The electric field here has an intensity of 100 dynes per statcoulomb." (The dyne is a unit of force; the statcoulomb is a unit of electric charge.) Should this remark be interpreted *literally*, as aiming to describe a certain entity (the electric field) that exists in just the way that tables and chairs exist? (To exist in this sense, a field need not be made of matter; "real" does not mean "material.") Or should a remark about the electric field be interpreted *non-literally*, so that its truth does not require that any such thing as a field *really exist*? For instance, perhaps the sentence "The

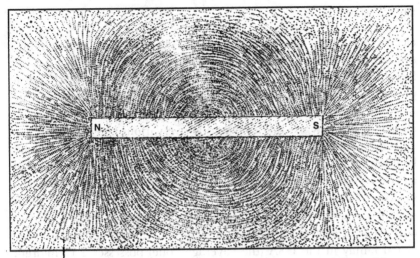

Figure P.1 Faraday's depiction of iron filings aligned along magnetic lines of force arising from a bar magnet. These lines, he writes,

> exist as much when there is no magnetic needle or crystal there [to detect them] as when there is; having an independent existence analogous to (though very different in nature from) a ray of light or heat, which, though it be present in a given space, and even occupies time in its transmission, is absolutely insensible to us by any means whilst it remains a ray, and is only made known through its effects. (1855: 323)

Magnetic lines of force had been plotted with a compass needle by Pierre de Maricourt about 1269, and iron filings had been used since at least the seventeenth century.

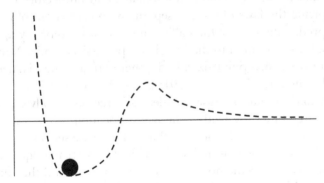

Figure P.2 The dotted line depicts the potential experienced by a proton as a function of distance from another proton. At large distances (greater than 2.5×10^{-13} cm), one proton electrically repels another, and this repulsion gradually weakens to arbitrarily small strength at long range. As the distance decreases, the strong nuclear force binding one proton to another predominates. But at distances below 5×10^{-14} cm, the strong nuclear force becomes very strongly repulsive. So protons in a nucleus are kept near but not too near each other. A proton is shown lying at the bottom of the "potential well."

electric field here has an intensity of 100 dynes per statcoulomb" simply means that *were* a charged particle to find itself here, then it *would* feel an electric force of 100 dynes for every statcoulomb of charge it possesses. In that case, the remark about the electric field may be true when no charged particle is here – even if there is no such thing as an electric field really existing over and above various bits of matter.

Alternatively, perhaps the remark about the electric field here should be interpreted (non-literally) as a convenient shorthand for the result of a certain complex calculation. In this respect, it would be similar to the sentence "The average family in the USA has 2.1 children." This sentence describes the result reached by a certain arithmetic procedure. It is true or false, but not in virtue of whether some particular actual family in the USA (the "average" one) possesses 2.1 children. The expression "the average family in the USA" does not refer to a family that exists on a par with your family and mine. Physics uses many concepts that seem similar to the concept of the average family in the USA, such as the concept of a system's *center of mass*, a kind of average location of all the mass in the system. To predict the behavior of a large collection of bodies, such as the atoms composing a billiard ball, it is often useful to imagine a single body the size of a geometric point, having a mass equal to the sum of the masses of the individual bodies in the collection, and located at the collection's center of mass. Rather than having to compute the billiard ball's trajectory by somehow applying the laws of motion separately to each atom of the ball, we can predict the path of the ball's center of mass simply by applying the same laws of motion to the imaginary point-sized body. Nevertheless, there is no such point-sized body; there are just the billiard ball's atoms (Dennett 1987: 52, 71; 1991: 27–9). Is the electric field also a useful fiction – useful because the electric force on a body equals the body's charge times the field at the body's location?

This was the kind of question that I found myself asking in my physics courses. The reason I can recall this particular question so vividly is because I remember how thrilled I was to find this question raised explicitly in my textbook, and then how confused and embarrassed I was to see it dismissed in these words:

Perhaps you still want to ask, what *is* an electric field? Is it something real, or is it merely a name for a factor in an equation which has to be multiplied by something else [a body's electric charge] to give the numerical value of the force [on the body] we measure in an experiment?

... [S]ince it works, it doesn't make any difference. That is not a frivo-
lous answer, but a serious one.   (Purcell 1965: 17)

Even after that somewhat defensive final sentence, Purcell offers no
justification for his "answer," and apparently he believes no justification necessary. Purcell's view seems to be that science is concerned
only with predicting what we will observe or measure in various circumstances (such as the force on a charged body) and so long as
electromagnetic theory is accurate for these purposes, it does not matter whether fields really exist over and above charged bodies. Indeed,
Purcell might well say that to try to figure out whether fields are real is
not only irrelevant to *scientific* purposes but pointless for *any* serious
purpose. We could never have a good reason for believing in the
reality or unreality of fields, so this is a merely "philosophical" issue,
impossible to investigate or to debate. This attitude is not unusual in
physics textbooks. Here is an excerpt from Richard Feynman's legend-
ary *Lectures on Physics*, from a section entitled "What are the fields?":

You may be saying: "... There are electric fields at every point in
space; then there are these 'laws' [that do not refer to fields, instead
telling us the force that one charge or current exerts directly on another, some distance away]. But what is *actually* happening? Why can't
you explain it, for instance, by whatever it *is* that goes between the
charges?" Well, it depends on your prejudices. ... The only sensible
question is what is the *most convenient* way to look at electrical effects.
Some people prefer to represent them as the interaction at a distance
of charges, and to use a complicated law. Others love the field
lines.   (Feynman et al. 1963: II, p.1-9)

The "only sensible question" – the only question that makes any sense
– is whether it is *"most convenient"* to think about electromagnetic interactions as involving fields, not whether this way of thinking about
them is *true*. These remarks made me feel very foolish for having
thought that it not only made sense to ask, but also was important to
answer the question "Are there really fields?"

I know *now* that it *is* important to answer this question and that (as
we will see) good reasons can be given for answering it a certain way.
Admittedly, the answer may make *no* difference to the way we should
use the concept of an electric field to predict the path of a charged
body. But the answer should make a *big* difference to our beliefs about
what the universe is like. This is the sort of topic that the philosophy of
physics investigates.

Feynman hints at one respect in which the reality of fields would make a difference. Consider a magnet and a compass some small distance away. The magnet is not in contact with the compass needle or with anything that is also touching the compass needle. Yet the magnet somehow manages to cause the needle to turn. This is weird: we tend to think that a cause must be in contact with its effect. How can the magnet affect the needle without something passing between them or their being in direct contact themselves? We will look closely at this question throughout the course of this book. For the moment, I shall say only that *if* fields are real, then the direct cause of the needle's motion is *local* to the needle: the magnetic field *where the needle is*. There is then no gap in space (or time) between the cause and its effect. But for the field to cause the needle to move, the field *must be real*, not merely a device we use to simplify our calculations.

The question of whether causes must be local to their effects is the main subject around which this book is organized. It is simultaneously a question of metaphysics – concerning the nature of cause and effect – and a question of physics. To pursue this question, we will have to grapple with many others: Are fields real? Is energy a real stuff that flows around or merely a calculational device? What about electric charge? Do material objects ever really come into contact? According to the theory of relativity, is energy actually matter, or is matter nothing but energy, or what? Does the "spooky action at a distance" in quantum mechanics undermine the idea that a cause must be local to its effect? Some of the issues with which we will have to wrestle involve general questions: about cause and effect, the way in which scientific theories are confirmed by evidence, the sense in which a theory can "unify" various phenomena, and the relation between a scientific theory and its philosophical interpretation. Other issues we will investigate are more closely associated with particular parts of physics.

I have tried to write the most introductory text there could be in the philosophy of physics that does not make undue sacrifices in either the philosophy or the physics. Accordingly, I have tried to presuppose a minimum of detailed prior experience with philosophy or physics. The issues on which I will concentrate might naturally strike any attentive reader of an introductory physics textbook. I have tried to motivate these issues in that spirit, sometimes using excerpts from physics textbooks to propel the discussion. Many of these issues not only were important in the development of classical physics, but also persist today in connection with the most up-to-date physical theories. This book does not pursue them nearly that far; in many respects, it merely scratches the surface. Nevertheless, it will leave you in a position to think rigorously about some of these issues for yourself. Accordingly, at the close of each chapter, I have included some questions for you to ponder. After all, it is much more fun to participate in these efforts than just to be a spectator.

It is also generally more fun to read books that stick their necks out once in a while by taking a stand than books that just give ever-lengthening collections of arguments, objections, and replies on all sides of various questions. So from time to time, you will find me defending some controversial philosophical view. You should take my conclusions to be provisional; by the time you read these words, I may even have changed my mind, for the arguments I will offer are not utterly conclusive. Perhaps that is for the best: I want you to read with an eye toward improving on what I say.

For that matter, probably none of the physical theories discussed in this book is completely true, though many of them may in some sense approximate the truth. For instance, although much of this book concerns the classical theory of electromagnetic fields, this theory cannot be married easily to charged quantum particles. And, of course, the coming years will undoubtedly bring new physical theories revealing the limitations of even our best current theories. So although the theories I shall examine are accurate enough for certain kinds of practical applications, their interpretation can give us only limited insight into reality. What, then, is the point of wrestling with their interpretation?

The answer is twofold. To begin with, many of the questions encountered in interpreting today's cutting-edge theories arose earlier in connection with the simpler theories investigated here. That is partly because classical electromagnetic field theory is the prototype for many current theories. But even if this were not the case, the theories I examine would still illustrate the interesting things that can happen when we take a theory from physics, a theory that seems straightforward enough when we are using it merely to predict our observations, and try to use it to describe reality.

Despite the continuing importance of the issues I shall discuss, several of them seem to have slid beneath the radar of current philosophy of physics, which is devoted largely to relativity, quantum mechanics, and beyond. This is understandable but unfortunate. An introduction that focuses primarily on the hottest topics in current physics not only makes little contact with the physics that most of us know most about, but also gives the impression that metaphysically speaking, everything was straightforward until relativity and quantum mechanics came along. This is a highly inaccurate picture. Philosophical puzzles arise from such familiar features of classical physics as collisions, potential energy, fields, and energy flow – puzzles with which nineteenth-century physicists were deeply engaged, as we shall see.

Furthermore, concentration on relativity and quantum mechanics may give the impression that philosophical work contributes nothing positive to science, but merely comes along after the real work has all been done, trying to cause trouble for some theory that scientists themselves find pretty satisfactory. After all, quantum mechanics was not developed in order to address *philosophical* problems with classical physics, and though philosophical grounds for discontent with quantum mechanics were pointed out shortly after its initial development, these concerns have as yet led to no amendment to the theory. On the other hand, Einstein arrived at relativity precisely by thinking about what he perceived to be problems in the *interpretation* of classical electromagnetic theory (as we will see in chapter 7). But these issues are neglected by texts that start with relativity and quantum mechanics, failing to look at classical electromagnetism; they emphasize only the *experimental* difficulties for classical physics that relativity addresses.

The problem with these texts is not merely their historical inaccuracy or their inability to account for Einstein's own remarks, such as this:

> What led me directly to the Special Theory of Relativity was the conviction that the electromotive force induced in a body in motion in a magnetic field was nothing else but an electric field.   (Holton 1973: 285)

Rather, the most harmful consequence of this approach is that it makes the questions investigated by philosophy of physics appear "merely philosophical" in a pejorative sense: marginal, detached from the concerns that actually drive innovation in physics. In fact, contrary to the remarks I quoted earlier from Purcell and Feynman, "philosophical" concerns have been (and continue to be) integral to progress in physics, as we will repeatedly see. I hope, then, that this book encourages students of physics and philosophy to continue their fruitful tradition of thinking hard about what reality has got to be like in order for our physical theories to succeed as well as they do in predicting our observations.

Enough polemic! Let's get started.

I owe thanks to Ann Baker, Jan Dyre Bjerknes, Hasok Chang, Bob Coburn, Arthur Fine, Tom Hankins, and Alex Rosenberg, who gave me very useful comments on some or all of the manuscript. Rob Deltete and David Keyt saved me from making some embarrassing mistakes. Three anonymous reviewers from Blackwell submitted detailed, encouraging, and constructively critical reports, for which I am very grateful. I especially thank Larry BonJour, who read the penultimate draft with tremendous care and offered good advice regarding both style and substance. The friendly atmosphere in my department, as well as the willingness of my philosophy of physics classes over the years to engage with this material, have also been of immense value.

My greatest debt is to my family – to my children, Rebecca and Abe, and especially to my wife, Dina – who manage somehow to love me even on those occasions when my philosophical preoccupations intrude on my better nature. I also thank Rebecca (age 9 years) for showing me how to use the computer for drawing the diagrams in this book.

Seattle, Washington   M.B.L.

1

What is Spatiotemporal Locality?

1 The Big Picture

Here's an old Russian/Jewish joke:

> The peasant Piotyr lives in the town of Kishinev early in the twentieth
> century. He has just witnessed the first demonstration in Kishinev of a
> shortwave radio receiver and transmitter. Someone has used the radio
> to converse with another person in Odessa, hundreds of miles away.
> Piotyr is astonished and asks his better-educated friend Ivan how this
> amazing invention works.
>   "Very simple," Ivan says. "Imagine a dog so large it stretches from
> Kishinev to Odessa. You step on the dog's tail in Kishinev and it barks
> in Odessa. Do you follow that?"
>   "I think so," Piotyr says, hesitantly.
>   "Good," Ivan says. "Now just remove the dog."   (Telushkin 1992: 60)

Of course, the joke is that Ivan doesn't succeed in making the radio
less mysterious to Piotyr. The radio used to be called the "wireless"
precisely because unlike a telegraph or telephone, there is no wire (or
apparently anything else) connecting the sender to the receiver. Under-
standably, Piotyr fails to grasp how the sender in Odessa manages
to cause his voice to be heard in Kishinev without anything like wires
to carry it there. Piotyr understands vaguely how your stepping on one
end of a dog causes a bark from the other end: something inside the
dog connects the tail to the mouth. But without the giant dog between
Odessa and Kishinev, there is no connection between the sender and
receiver, and so the original mystery remains: how does the cause in

Odessa have an effect in Kishinev without anything to carry its influence across the space in between?

Like Piotyr, we expect an effect to be located at the site of its cause. For example, suppose that a window shatters. The cause might have been a certain baseball's colliding with it. But suppose the ball was never anywhere near the window. Then the ball could not have had anything to do with the shattering. A cause must be *local* to its effect.

Of course, your stepping on a dog's tail causes the dog to bark despite the dog's mouth being some distance away from its tail. But this is understandable: a *chain* of causes and effects connects your stepping on the dog's tail to the dog's mouth barking. Your stepping on the tail has the *direct* effect of compressing certain skin cells. This compression stimulates nerve cells, which influence cells in the dog's brain, which excite nerve cells projecting to the dog's mouth, throat, and chest, which cause certain muscles to move, which ultimately produces a bark. The *direct* effect of each cause in the chain occurs locally to that cause. Through this chain, your stepping on the dog's tail has an *indirect* effect some distance away.

We expect an effect to be local to its cause both in space and time. For example, the cause of the window's shattering must have been happening when the window began to shatter. Of course, we should again distinguish direct from indirect causes. For example, suppose that lightning strikes a tree, ultimately causing a fire. The fire may not ignite until some minutes after the lightning strikes. But the fire is not a *direct* effect of the lightning strike. Rather, the lightning directly caused some chemical changes in the wood that it struck. These changes ultimately produced the fire. If, a moment after it was struck by lightning, the tree had been no different from the way it was a moment before, then (we think) the lightning could not have caused the fire that ignited later. A cause cannot make itself felt a few minutes after it occurs without having effects at every moment in between, just as (Piotyr presumed) a cause cannot have an influence at some distant location without having effects at all of the locations in between. I will call this important idea *spatiotemporal locality*.

"Action by contact," as when two billiard balls collide, is an obvious illustration of spatiotemporal locality. (However, even "action by contact" may raise serious puzzles, as we will see shortly.) "Action at a distance," as when two bodies separated by a gap in space affect each other gravitationally, electrically, or magnetically, seems to violate spatiotemporal locality. But these interactions are local if *fields* are real;

the gravitational, electric, and magnetic forces felt by a given body would then be *directly* caused by the corresponding fields *at that body's location* rather than by some *distant* body.

What *reasons* are there for believing fields to be real? (We cannot justify this belief on the grounds that gravity and electromagnetism would then obey locality – unless we *already* had a good reason for believing in locality!) As we will see, many philosophers and physicists worked to ascertain whether fields are real, leading to classical electromagnetic theory and the special theory of relativity. Furthermore, the reality of fields is bound up with many other fascinating issues in the interpretation of physical theories, including the character of energy and matter as real kinds of "stuff," the sense in which electricity has been *unified* with magnetism (and energy with mass), the apparent need for fields to be real in order for there to be explanations of certain facts, and the distinction between a scientific theory and its philosophical interpretation. By exploring issues arising from spatiotemporal locality, we will run into many (though not all!) of the most fundamental and historically significant questions in the philosophy of physics. This is the trail we will follow throughout this book.

Our first step must be to clarify "spatiotemporal locality." That's the task of this chapter.

2 Causal Relations between Events

Can there be space or time separating a cause from its direct effects, or must a cause be local to its effects? I will presume that this question makes sense. But it makes sense only if a cause and its effects have locations in space and time. Otherwise, we can't ask whether they must be near each other.

For causes to have locations in space and time, a cause needn't be confined to a single *point* in space and *instant* in time. A cause may occur throughout some time period or over an entire volume in space. For example, suppose that the shattering of a certain window was caused by a baseball's colliding with it. If the window was three feet wide, two feet high, and a quarter-inch thick, then its shattering took place in the entire region of space that the window was filling.

I just referred to the shattering as "taking place." Its cause, the collision between a baseball and the window, is also something that "occurs." Accordingly, it seems natural to call the collision and shattering *events*.

An event is a particular thing that happens somewhere and somewhen. Since it makes sense to ask whether a cause must be local in space and time to its effects, causes and effects must be events.

Here's another example: my pressing steadily but gently on my car's brake pedal causes the car to slow down gradually to a stop. The cause and its effect occur throughout a certain temporal *interval* a few seconds long. Since the car is moving while it is slowing down to a stop, the cause (my pressing on the brake) has a different location at different moments. The same applies to the effect: at a given moment, the car's slowing down to a stop is occurring wherever the car is. Since the car is moving, the event of the car's slowing down to a stop *moves* while it occurs.

Certain events would not ordinarily be described as "occurring" or, indeed, as "causes." For example, the lightning strike is *only one* of the events that conspire to cause the fire's igniting. It is not the *complete* cause. One of the other causes that combines with the lightning strike to form *a complete set of the fire's causes* is the tree's being dry. Another is the tree's being surrounded by oxygen. Ordinarily, if someone asked us why the tree caught fire, we might mention only the lightning strike; the person who asked presumably already knew that the tree was dry and surrounded by oxygen. Perhaps she mainly wanted to know why it caught fire *just when it did* rather than at some earlier moment or never. Only the lightning strike explains why; its occurrence is the only thing making the moment at which the tree caught fire relevantly different from other recent moments. Nevertheless, the tree's being dry and surrounded by oxygen helped to cause the fire to ignite. They are *causally relevant*, and that is all I mean by referring to them as "causes" of the fire.[1] We would not usually refer to them as "events" that "occur." Yet that is what they are.[2]

To help clarify what an event is, let's contrast events with facts. An event occurs somewhere somewhen; there are times when and places where it is not occurring. On the other hand, a fact holds everywhere and everywhen. For example, if it is now (at 1:45 p.m. on October 22, 1999) a fact that I am in my office at 1:45 p.m. on October 22, 1999, then in 1543 in Florence it was a fact that I am in my office at 1:45 p.m. on October 22, 1999, and this fact will still be a fact on Mars in the twenty-second century (see box 1.1). Of course, in Florence in 1543, no one yet *knew* this fact about me!

Perhaps a fact is just a truth.[3] It is true that I am in my office at 1:45 p.m. on October 22, 1999. But the window's shattering, an event, is

Box 1.1
The openness of the future

Some philosophers reject this view of facts as conflicting with the future's "openness": If it is already true *now* that humanity will colonize Mars next century, then the future is settled, so our current actions cannot affect it (Prior 1953). This worry is misplaced. If humanity colonizes Mars next century, our current efforts will have made humanity's later success possible. Success was not inevitable; were we now not exploring space, it would be true neither now nor later that humanity colonizes Mars in the twenty-second century.

not *true*. It's not false either. It is not the *sort* of thing that *could* be true or false. "The window's shattering" is like "My house" in that it would be a "category mistake" to call it true (or false). The window's shattering *happens*; it *takes place*. Of course, it is true *that* the window shatters. That the window shatters is a fact. A fact *involves* various particular things, such as a window. An event *is* a particular thing, and there are facts about an event just as there are facts about a window. Here is a fact about an event, the window's shattering: it was caused by the baseball's colliding with the window.

Of course, an event and a house can both be particular things and nevertheless differ in important ways. A house is made of matter, has mass, and is a certain color, whereas none of these properties can be possessed by any event. Furthermore, two events can occur at the same place and time (Lewis 1986: 245). For example, in Newtonian physics, when a body feels a certain force and is thereby accelerated, the body's feeling that force is an event occurring at the same place and time as the body's accelerating in response. On the other hand, two objects made of matter cannot occupy the same place at the same time. (Or so we are inclined to think – but see the next section!)

Although an event is less tangible than a house, they are alike in certain basic respects simply because they are both particular things. For instance, many different expressions refer to the same house: "the house next door," "the Jones's house," and "the building that I am thinking about right now." Likewise, many different expressions pick out the same event: "the shattering of the window," "the cause of that sound of glass breaking," and "the first event I told my wife about when she came home."

Facts do not work this way. Even though Marc Lange is the author of this book, the fact that Marc Lange is in his office is not the same as the fact that the author of this book is in his office. The second fact requires the cooperation of this book; the first fact does not. The second fact could have held without the first fact's holding. (Try to imagine how!) One fact (that Marc Lange is in his office) can *entail* another (that Marc Lange has an office) in the sense that it would be a *contradiction* for the latter to be false and the former true. The fact that Marc Lange is in his office stands in different inferential relations from the fact that the author of this book is in his office. (For example, only the former implies that Marc Lange has an office.) Hence, they are different facts. Events cannot stand in inferential relations at all. No sentence's truth is guaranteed by the truth of "The window's shattering" because "The window's shattering" is not the sort of thing that *can* be true.

A causal relation sometimes seems like a relation between *facts*, as when we say "The window shattered because the baseball collided with it." Of course, we could instead have said "The shattering of the window was caused by the baseball's colliding with it," portraying *events* standing in a causal relation. Perhaps causal relations hold among facts and also among events (Bennett 1988). Even if this is so, our interest lies in spatiotemporal locality, and therefore we must look at causal relations among events, since facts have no spatiotemporal locations. On the other hand, perhaps events *alone* stand in causal relations (Davidson 1980). Then the "because" relation in "The window shattered because the baseball collided with it" cannot be causal, even though it is *explanatory*: the fact that the baseball collided with it explains why the window shattered (rather than remaining intact).

Suppose the same event is picked out by "the lightning's striking the tree" and "the event that I am thinking about right now." So the event that I am thinking about right now caused the fire's igniting. Of course, my thinking about it right now did not enable it to cause the fire! This just goes to show that the properties an event possesses by which we identify it on some occasion need not be the properties that make it another event's cause. The lightning strike's electrical properties are the only features it possesses intrinsically that are *causally relevant* to the fire's igniting. (I will say more about "intrinsic" properties in section 5.)

3   Action by Contact

The idea behind spatiotemporal locality is that there can be no gap in space or time between a cause and its direct effects. It would strike many of us as mysterious – magical, even – for two billiard balls to interact without touching. If a person in Odessa talks to a person in Kishinev – if, as we say, they are "in touch" or "in contact" with each other – then we expect them to be touching or for there to be another object between them that they are both touching.

"Action by contact" has often been considered unproblematic. Here, for instance, is William Thomson (Lord Kelvin, the Scots mathematician and physicist) in 1884:

> It seems to me that the test of "Do we or do we not understand a particular point in physics?" is "Can we make a mechanical model of it?"   (1987: 111, also 206)

But as the following discussion shows, causation by mechanical impact – touching – is not as straightforward as it might appear.

Consider points A and B in space (or instants A and B in time). They are separated by some distance (or span of time), which must contain other points (or instants). Suppose we select one: C. Now we can repeat the procedure: between A and C, there is some separation, within which lies another point (or instant): D, and so on. Hence, between A and B, there are infinitely many points (or instants). In short, classical physics assumes space and time to be "dense": between any two points in space (or moments in time), there is another. Closely related is the idea that classical physics treats trajectories as "continuous." For example, if between times $t = 0$ and $t = 1$, a body moves from the spot with coordinates $(0,0,0)$ to the spot with coordinates $(1,1,1)$, then for any real number n between 0 and 1, there is an instant of time $t = n$ and at that instant, the body has a definite location (coordinates (a,b,c), where a, b, and c are real numbers).

Imagine, then, that the basic constituents of matter are *point* particles. They have no length or width; each has zero volume. The only way they can touch is for them to occupy *the same point* in space. (They cannot touch by occupying *neighboring* points since there are no neighboring points; between any two points, there are infinitely many others.) But if two particles can occupy the same point at the same instant, then they cannot be hard, solid, impenetrable bodies. Indeed, since they can get inside each other, it is difficult to understand why they bounce off each other; when they collide, they might just as well pass right through each other! This seems bizarre: the whole point of matter is its solidity – that one bit of matter excludes every other from the space it occupies.

To avoid this puzzling result, imagine the two bodies not to be geometric points. Take two billiard balls colliding. The causal relation between them should definitely qualify as spatiotemporally local – as the *opposite* of action at a distance!

Let each billiard ball be a sphere of radius r. Each ball occupies all of the points at a distance from its center less than or equal to r. What happens when two balls touch? Their centers are separated by a distance of $2r$. A point located exactly between the two balls, at a distance r from the center of each, is where the balls touch when they collide: the "point of osculation" (literally: the kissing point). *Both* balls occupy that point, since (to repeat) each ball occupies all of the points at a distance from its center less than *or equal to r*. We have re-created the problem we faced with the point particles: two bodies occupying the same location at the same moment. This *seems* problematic, though the Irish physicist George Francis FitzGerald (1851–1901, perhaps best remembered for the "Lorentz–FitzGerald contraction" later incorporated by Einstein into relativity theory) wrote nonchalantly:

> When rigid bodies act on one another by non-slipping contact, certainly the coordinates of the point of contact are common to the two systems. (1895: 285)

Perhaps FitzGerald is correct. The northern and southern hemispheres of Earth are in contact and together constitute the entire Earth, leaving nothing out. There is no reason for the Equator to belong to one hemisphere rather than the other. So the Equator must belong to *both*. On the other hand, the Equator is merely an artificial, conventional boundary drawn *within* a single body. Unlike two billiard balls, the two hemispheres are not distinct bodies in contact.

The problem of overlapping bodies would not arise if a billiard ball occupied only the points at a distance *less than r* from its center. This conception might be inspired by a question famously asked in an 1893 paper by the American philosopher, physicist, and mathematician Charles Sanders Peirce (1839–1914). Regarding a solid black circle on a white piece of paper, Peirce (1933: 98) asked: Does the boundary line belong to the circle or the background? Are the points on this line black or white? Remember, a point has no width so it cannot be half black and half white. Moreover, a line in geometry is only one point wide, so it cannot be that each point on the outer half of the line is entirely white and each point on the inner half is entirely black.

Apparently, no view is so weird that it has never been defended in an interesting way by some philosopher. (That, to me, is one of the joys and strengths of serious philosophizing.) This case is no exception: Franz Brentano (1838–1917, Austrian philosopher and psychologist) suggested (1988: 41) that at the boundary line, a black line and a white line coincide; each point on the line is entirely black *and* entirely white!

At least, there is no obvious reason to award the boundary line to the circle rather than to the background; each seems to have as good a claim on it as the other. We might, then, be inclined to say that a point on the line is neither black nor white. On this view, neither the solid circle nor the background includes the boundary, though each region extends right up to the line. Analogously, perhaps the points at a distance of exactly r from the billiard ball's center do not belong to the ball, though the ball extends right up to them (Kant 1985: 63).

In that case, however, two billiard balls in contact must have a gap at least a point wide separating them. The two balls would be touching and yet any continuous path in space from one to the other must cross a point occupied by neither. This doesn't seem like contact at all![4] Furthermore, if the balls could interact across a gap of one point, then why not across a gap of two points, or even two miles? Action "by contact" would then seem as "magical" as action at a distance!

This problem does not arise if *one* of the colliding balls occupies all of the points at a distance from its center *less than or equal to* r, whereas *the other* occupies only those points *less than* r from its center. When the balls touch, they are like two parts of a single body: there is continuity without overlap. In defending this view, Bernard Bolzano (Czech philosopher and mathematician, 1781–1848) defined

> the contact of two bodies as taking place when the extreme atoms of the one, . . . together with certain atoms of the other, form a continuous extension. (1950: 168)

On this view, the answer to Peirce's question is that the boundary line belongs either to the black circle or to the white background (but not to both). Perhaps if we put black paint on white paper, then the boundary line belongs to the circle, whereas if we paint a white surround on a paper that was already black, then the line belongs to the background.

But applied to billiard balls, this view is unattractive. Brentano (1988: 146) called it "monstrous"! Surely all billiard balls of radius r are alike in either including or excluding the points at radius r. If some are "inclusive" and others "exclusive," then (on Bolzano's view of what contact requires) an "inclusive" ball cannot collide with another "inclusive" ball (since they cannot occupy the same point at the same time) and an "exclusive" ball cannot collide with another "exclusive" ball (since there is at least a point in between). Billiards would be a very difficult game.

As I warned you, action by contact is not straightforward. I have just surveyed three ways of understanding two colliding billiard balls: as sharing a point, as separated by a point, and as momentarily forming a continuous body. Each view encountered severe difficulties – perhaps not severe enough to prove it false, but enough to be worrisome. Accordingly, let's rethink our picture of collisions.

To gain some perspective on our problem, let's examine instead a single billiard ball moving in empty space. Suppose its speed at time $t = 0$ was 5 meters per second, and that from $t = 0$ to $t = 5$, it feels no forces. By Newton's laws of motion, it undergoes no acceleration and so continues moving at 5 m/s through $t = 5$. What, then, caused the body at $t = 3$ to be moving at 5 m/s? By spatiotemporal locality, the *direct* causes cannot be separated by any gap in space and time from the effect. Given Newton's laws, the direct causes must be the body's already moving at 5 m/s at some moment and its feeling no forces between that moment and $t = 3$.[5] But it's already moving at 5 m/s *when?*

The body's moving at 5 m/s at $t = 3$ cannot be a cause of the body's moving at 5 m/s at $t = 3$, since an event cannot be one of its own causes. And according to locality, the body's moving at 5 m/s at some moment *earlier* than $t = 3$ (say, at $t = 2.5$) cannot be a *direct* cause of its moving at 5 m/s at $t = 3$, since this would create a gap in time (of 0.5 second) between the effect and its direct cause. This is like the problem we encountered in trying to apply locality to two colliding billiard balls. There we faced the unwholesome choice between the balls overlapping at a point or not overlapping and so being separated by at least a point. With the isolated billiard ball, we face an analogous unwholesome choice: its moving at 5 m/s at $t = 3$ is caused either by itself or by its speed at a moment separated by a gap from $t = 3$.

These problems all arise from space and time being "dense." Between any two points in space (or time), there are infinitely many other points: a gap. Apparently, for a cause to be local to its direct effect – for there to be no gap between them – the cause must share a point in space (and moment in time) with its direct effect. But this seems implausible: it is problematic for two bodies to overlap, and an event cannot be its own cause.

A possible solution is to rethink the distinction between direct and indirect causes. Since time is dense, there is no *unique* "direct" cause of the isolated ball's moving at 5 m/s at $t = 3$. There is no moment where $t = 3$ is *the next moment*. So there is no particular moment where the body's moving at 5 m/s at that moment (and the body's feeling no force from then to $t = 3$) is *the direct cause* of the body's moving at 5 m/s at $t = 3$.

However, within *any* interval prior to $t = 3$, *no matter how short*, there is a complete set of causes of the body's moving at 5 m/s at $t = 3$. For example, within the interval between $t = 2.9$ and $t = 3$, there is the body's moving at 5 m/s at $t = 2.99$ and its feeling no forces between $t = 2.99$ and $t = 3$ inclusive. Within the narrower interval between $t = 2.999$ and $t = 3$, there is another complete set of causes: the body's moving at 5 m/s at $t = 2.9999$ and its feeling no forces between $t = 2.9999$ and $t = 3$ inclusive (see box 1.2). There is a complete set of causes *arbitrarily near* in space and time to the effect. Therefore, I suggest, there is no "gap" between cause and effect, satisfying spatiotemporal locality.

It might be objected: an arbitrarily small gap between cause and effect is *still a gap*. There *is* a separation between $t = 2.99$ and $t = 3$, and

---

Box 1.2
A complete set of causes

A "complete" set of causes of an event is like *one* link in the causal chain leading to that event. It is "complete" in the sense that it constitutes a complete link in that chain, not in the sense that it includes all of the event's causes (that is, all of the other links). The same event may therefore have *many* complete sets of causes. An analogy: My four grandparents might (loosely) be termed a "complete" set of my causes, even though there are other links in the chain (such as the link consisting of my parents). Three of my grandparents, without the fourth, would not be a complete set of my causes.

there is a smaller separation between $t = 2.9999$ and $t = 3$, and as we extend the sequence further, there will always be some finite separation. Accordingly, Kline and Matheson (1987) insist that contact requires no separation at all.

But then spatiotemporal locality is violated by exactly the cases that we would initially have considered prime examples. Rather than argue about what "contact" really means, I suggest that we define "spatiotemporal locality" in whatever manner best fits our initial ideas about the sorts of cases that should count as exemplifying it. We can then investigate whether this sort of locality holds in every actual case.

What does this approach say about billiard balls colliding? If each ball occupies only the points less than r from its center, then on this approach, the balls are "touching" when they are separated by only one point. It may appear contradictory for contact to involve a gap, even of one point. But if a "gap" must be a *region* of space – a finite non-zero volume – then there is *no gap* between the balls. If a "distance" must be some *finite* quantity, then two balls separated by a single point have no distance between them and so fail to involve action at a distance. That two bodies can be "aware" of each other across an unoccupied point does *not* suggest that they can "know" about each other across a gap of two points – which is a gap of infinitely many points, because space is dense. A "gap" of one point is *qualitatively different* from anything larger: it is not a *region* of space. Again, I recommend that "spatiotemporal locality" be defined to capture the cases to which we think it ought to apply most easily. Otherwise, the issue of whether "spatiotemporal locality" holds becomes quite different from the issue we set out to examine.

Of course, I have given no reason to believe that a billiard ball occupies only the points at distances less than r from its center. If instead each ball occupies all of the points at distances less than *or equal to r*, then perhaps we should join FitzGerald in allowing two bodies to share a point, but prohibit them from sharing a region. This allows spatiotemporal locality to apply to "action by contact." Does the "impenetrability" of matter preclude bodies from overlapping in a finite non-zero volume but permit them to overlap at a single point?

After we have introduced fields of force, we may be able to picture the "collision" between two balls as involving repulsive fields driving the balls apart. Each ball interacts only with the field "surrounding" the other ball. A "collision," then, is just a close approach between the two balls – close enough that each ball penetrates deeply into the other's field, to where that field becomes very strong. Spatiotemporal locality is guaranteed by each ball's interacting only with the field *at its location*. (These are "coming attractions.")

4 Spatial, Temporal, and Spatiotemporal Locality Defined

Our discussion of locality began with the thought that two things directly "in touch" with each other must be touching, so that nothing can be squeezed between them. Given the density of space and time, this became the idea that no *region* of space or *interval* of time can separate them – that nothing *extended* in space or time can fit between them. When no *unique* set of *direct* causes exists, locality says that a cause can have an influence at some distant location (or at some later moment) only by having effects at every location (or during every moment) in between. In other words, no extended region of space (or interval of time) exists around the effect that does not contain a complete set of its causes.

To sharpen this notion of spatiotemporal locality, let's distinguish "temporal locality" and "spatial locality."

Here's what "temporal locality" means. Take an event E. Choose a finite, non-zero amount of time τ: perhaps let $\tau = 1$ sec, or even let $\tau = 0.0000001$ sec. Temporal locality says that no matter how short τ is, the moments within τ of E contain a complete set of E's causes.[6] In other words, E is not surrounded by some finite non-zero gap (time τ in length) during which *no* complete set of E's causes occurs. More precisely:

> *Temporal locality*: For any event E and for any finite temporal interval $\tau > 0$, no matter how short, there is a complete set of E's causes such that for each event C (a cause) in this set, there is a moment at which it occurs that is separated by an interval no greater than τ from a moment at which E occurs.[7]

C and E may each last for more than a single moment. Temporal locality requires that *at least one* of the perhaps many moments at which C occurs be within τ of *at least one* of the perhaps many moments at which E occurs. For simplicity, however, consider the isolated billiard ball case, where E (the ball's moving at 5 m/s at $t = 3$) lasts for only a moment. If $\tau = 0.1$ second, then here is a complete set of causes where each event C in the set occurs at a moment no more than τ away from E: the set consisting of the ball's moving at 5 m/s at $t = 2.99$ and the ball's feeling no forces between $t = 2.99$ and $t = 3$. No matter how small τ becomes (so long as τ remains non-zero), there remains such a set.

Likewise, "spatial locality" means that E is not surrounded by some finite gap in space, extending out to a non-zero distance δ, in which *no* complete set of E's causes occurs. More precisely:

> *Spatial locality*:   For any event E and for any finite distance $\delta > 0$, no matter how small, there is a complete set of causes of E such that for each event C in this set, there is a location at which it occurs that is separated by a distance no greater than δ from a location at which E occurs.

C and E may occupy extended regions of space. Spatial locality requires that *at least one* of the perhaps many points at which C occurs be within δ of *at least one* of the perhaps many points at which E occurs. (In fact, given two events, C_1 and C_2, that form a complete set of E's causes, spatial locality is satisfied if there is a point where E occurs that is within δ of C_1 and *another* point where E occurs that is within δ of C_2, even if there is no *single* E-point that is within δ of *both* C_1 and C_2.)

Spatiotemporal locality is not simply spatial locality plus temporal locality. That is because spatial locality requires only that there be a point in space at which C occurs *sometime* that is separated by a distance no greater than δ from a point in space at which E occurs *sometime*. This imposes no restriction at all on how close those times have to be. It could be that we cannot bring these times close together without disrupting the spatial closeness. Then *spatiotemporal* locality would be violated.

In other words: Suppose that spatial locality holds of E. Consider a complete set of E's causes where each event C in that set occurs *sometime* within a distance $\delta > 0$ of some point at which E occurs *sometime*. Suppose that every C in this set is separated in time from E by at least the interval $\tau > 0$. That is, every C in this set occurs *outside* the small interval τ around E. Suppose further that all this is also true for *every* complete set of E's causes inside δ. So to get causes within τ of the effect, as temporal locality demands, we cannot use causes within δ. Nevertheless, it may be the case that no matter how small $\tau > 0$ is, there is a complete set of E's causes within τ of E. But once τ falls beneath a certain value, every C in any complete set of E's causes occurring within τ of E is never within δ of a point where E occurs sometime. Then although *both* temporal locality *and* spatial locality hold of E, the sets of E's causes that make spatial locality hold are not the same as the sets of E's causes that make temporal locality hold. Intuitively, *spatiotemporal locality* then fails to hold, since it requires that we be able to diminish arbitrarily the spatial *and* temporal gaps *together.*

Spatiotemporal locality: For any event E, any finite temporal interval τ > 0, and any finite distance δ > 0, there is a complete set of causes of E such that for each event C in this set, there is a location at which it occurs that is separated by a distance no greater than δ from a location at which E occurs, *and* there is a moment at which C occurs *at the former location* that is separated by an interval no greater than τ from a moment at which E occurs *at the latter location.*

Spatiotemporal locality requires spatial locality and temporal locality, but spatial locality and temporal locality could both hold even if spatiotemporal locality is violated.

Let us look at some examples. Take gravity according to Newton's law: A point of mass M exerts on a point of mass m, at a distance r, an attractive force of magnitude GMm/r^2. *Temporal* locality is satisfied since the effect (the gravitational force at time t impressed by one body on another) has a complete set of causes *simultaneous* to it (namely, the masses and separation of the bodies *at t*). However, *spatial* locality is violated; this is action at a distance in its purest form. For example, the point masses composing the Sun cause forces on the point masses composing the Earth across a gap of some 93 million miles. On this interpretation of the physics, there are no causally relevant events between the Sun and the Earth. (This will change when we introduce fields in the next chapter.)

Alternatively, imagine that the Sun's gravitational influence is *not* felt *instantly* on Earth. Suppose there would be a 500-second delay between a change in the Sun's mass and the corresponding change in the force on Earth. That is how long sunlight takes to reach Earth. Sunlight arrives at Earth only after first being one mile from the Sun, shortly later two miles, then three, and so on. But imagine that there are no events between the Sun and Earth that are causally relevant to the gravitational force exerted by the Sun on the Earth. Rather, the

---

Box 1.3
Other senses of locality

In the philosophy of physics, "locality" is used in many different senses besides the senses given here (Earman 1987). It is sometimes used to mean that C can be a cause of E only if the separation in space between C and E, divided by their separation in time, is finite (Reichenbach 1958: 131–2) – in other words, only if the speed at which you would have to travel to get from C to E is finite (even if high). The 500-second delay between the cause on the Sun and the effect on the Earth would not constitute a violation of locality in this sense. Alternatively, "locality" sometimes means that causal influences propagate no faster than the speed of light (in a vacuum), so that C cannot be causally relevant to E if they are "spacelike separated" – that is, if one would have to exceed the speed of light in order to reach E from C. These notions are different from "spatiotemporal locality." For example, a field version of Newtonian gravitation satisfies spatiotemporal locality (because of the field, there are no gaps) but causal influences propagate infinitely fast (that is, instantaneously throughout the universe). Conversely, if C and E are located at the same place at different times, C preceding E, then they are guaranteed to be "local" in the sense that an influence could get from C to E without traveling faster than the speed of light in a vacuum (it would not have to travel at all). But spatiotemporal locality could be violated (as in the next example in the main text). Other senses of "locality" have nothing directly to do with causal relations, such as the idea (discussed in chapter 9) that a system's intrinsic properties are determined by the properties of its parts, and their intrinsic properties are determined by the properties of their parts, and so on all the way down to the properties at points, the most "localized" properties.

---

latest cause of a certain gravitational force on Earth is an event occurring on the Sun 500 seconds earlier. Then not only spatial locality but also temporal locality would be violated (but see box 1.3).

The next example is entirely fictional. Suppose a body passes through a given location, leaving no trace behind it. One hour later, another body passes through there and is affected by the first body's having been there one hour earlier. Remember, nothing was left behind by the first body to be encountered later by the second. So spatial locality applies whereas temporal locality is violated.

Of course, this example may strike you as wacky: How could the first body's passing through affect the second body one hour later unless the first body left a residue behind? In that case, however, the residue would be a *local* cause of the effect on the second body. If the original example seems bizarre to you, that's perfectly fine; your reaction merely reflects how plausible locality seems to you. Nevertheless, locality appears to be violated by gravitational and electromagnetic interactions. We will look into this in subsequent chapters.

5 Intrinsic Properties and Noncausal Connections

Return to the case we just discussed, which was meant to demonstrate that temporal locality can be violated while spatial locality is satisfied. Here's a strategy you might consider for making this case satisfy temporal locality too. Notice that just as the second body passes through the given location, the following event occurs there: its becoming one hour since the first body passed through. This is a peculiar way to identify an event; I have mentioned nothing happening except for a certain amount of time having passed since the first body came through. This event occurs at the same moment as the effect on the second body. So if this event is causally relevant to the effect, then temporal locality is upheld! But this event must not qualify as causally relevant, else temporal locality will hold *no matter what*! It will hold *trivially*, no longer allowing us to draw the kind of interesting distinction that we wanted to draw between action by contact and action at a spatiotemporal distance.

A similar tactic would trivialize spatial locality. It is supposed to be violated by gravity operating according to Newton's law. Some gravitational forces on Earth are caused by the matter in a certain region of the Sun having mass M. This cause occurs in that region of the Sun, far away from Earth. But at the same moment, a certain event occurs *on Earth* (spatially *local* to the effect). Here it is: Earth at that moment belonging to a universe where a certain region of the Sun has mass M at that moment! (Compare Shoemaker 1969: 378–80.)

Obviously, if our locality principles are going to have any bite, we must somehow conclude that these events cannot be causally relevant. Let's try to find a good reason for doing so. Take the property of being red. Suppose it is possessed in a certain place at a certain time. For example, a shirt there and then is red. So redness is "instantiated" there and then. An event involves certain properties being instantiated in a certain region (or point) of space at a certain interval (or instant) of time, which constitute the event's "spatiotemporal location."

Take the property of containing 5 kilograms of matter. For it to be instantiated at a certain spatiotemporal location, something must be going on *at that location* – namely, 5 kg of matter must be there. That this property is instantiated at a given spatiotemporal location imposes some constraints on *that location*, but seems to impose no logical constraints on what properties may be instantiated at *other* locations. For example, no *contradiction* would ensue if 5 kg of matter occupied a given point in space at a given moment, but at every other moment in the universe's past and future history, there was less than 5 kg of matter in the entire universe. Of course, this scenario may violate some *law of nature* – for instance, a law that the total quantity of matter in the universe must remain constant (the conservation of matter). But presumably, a violation of this law would involve no *logical contradiction* (unlike "There exists a round square," which is logically contradictory). There being 5 kg of matter here now seems to impose no *logical* restrictions on what properties can be instantiated elsewhere elsewhen.

In other words, containing 5 kg of matter is an *intrinsic* property of a region of space. Roughly speaking, an intrinsic property of a thing requires something only *of that thing*; its intrinsic properties depend only on itself, not on its relations to other things. (Of course, a body's intrinsic properties can be *changed* by other things' effects on it, but the properties are possessed by the body entirely in virtue of what the body is like.) A body's mass, electric charge, and shape are among its intrinsic properties (at least in classical physics).

In contrast, a typical non-intrinsic ("extrinsic") property is the property of standing in a certain *relation* to something else. For example, the property of being 5 meters away from 5 kg of matter is *not* an intrinsic property: its instantiation here now requires the cooperation of other spatiotemporal regions. Obviously, for this property to be instantiated here now, there must be 5 kg of matter 5 meters away.

A body's intrinsic properties do not depend on other bodies (or anything else external to the body). For example, suppose I now possess the property of being the tallest person within 100 meters. This is not one of my intrinsic properties, since my possessing it depends on certain other people (who are taller than I am) not now being within 100 m. In other words, my now being the tallest person within 100 m depends not just on what is going on *at my location*, but also on what is happening in the rest of the region 100 m around me. In contrast, my being 148 cm tall is one of my intrinsic properties; it depends only on me. If different people come within 100 m of me but I am left alone, I will still be 148 cm tall. However, I may no longer be the tallest person within 100 m.

Having said this, I must now take some of it back. Although the property of containing 5 kg of matter is *supposed* to qualify as intrinsic, its instantiation here now does *sort of* logically require the cooperation of other spatiotemporal locations: for this property to be instantiated here now, there is a property that any point 5 meters away must instantiate now. Can you think of it? Of course: the property of being 5 m away from 5 kg of matter! However, this property is not intrinsic. So we have arrived at this idea: A property is "intrinsic" if and only if its instantiation at one spatiotemporal location (a point or a region) puts no logical restrictions on the *intrinsic* properties instantiated at locations that do not overlap the first location.

Of course, this idea cannot be used to *define* "intrinsic property" since "intrinsic" appears on *both* sides of the "if and only if." Nevertheless, it is a useful idea; let's refine it. Consider the property of containing 5 kg of matter *or* being 5 m away from 5 kg of matter. This property is *not* intrinsic, yet its instantiation here now imposes no logical restrictions on the intrinsic properties instantiated elsewhere elsewhen. (For example, that this property is instantiated here now does *not* logically require that there be 5 kg of matter 5 m away; the property is instantiated here now if there is 5 kg of matter here now, regardless of what is going on elsewhere.) This example suggests that for a property to be intrinsic, it is necessary *but not sufficient* that its instantiation here now impose no logical restrictions on the intrinsic properties instantiated elsewhere elsewhen (Lewis and Langton 1998).

Now take the property of having a brother (one who is currently living, I mean). Jones's here now having a brother logically restricts the intrinsic properties instantiated in other regions of space-time: for example, there must now somewhere be another human being (namely, Jones's brother). Consider, then, the property of *not* having a brother (alive right now) – being "brotherless." Since *having* a brother is a non-intrinsic property, being *brotherless* must be non-intrinsic. Yet Smith's being brotherless here now imposes no logical restrictions on the intrinsic properties instantiated elsewhere elsewhen. (Smith can be brotherless whether or not there are other human beings around.) So let's amend our proposal to read: if property P is intrinsic, then no logical restrictions on the intrinsic properties instantiated at other places or times are imposed *either* by P's being instantiated in a given spatiotemporal region *or* by not-P's being instantiated there. So if containing 5 kg of matter is an intrinsic property, then being 5 m away from 5 kg of matter is *not* an intrinsic property, since the latter's instantiation here now logically requires the former's instantiation now at the circle formed by the points 5 meters away.

We can use the notion of an intrinsic property to explain why the events threatening to trivialize our definition of "spatiotemporal locality" cannot be causes. Consider again a gravitational force at time t on a chunk of the Earth having mass M. One of its causes (according to Newton's law) is a region of the Sun (a distance r away) possessing mass m at t. At the place on Earth where this gravitational force occurs at t, there also occur at t various events possessing this *non-intrinsic* property: occurring a distance r away from a region containing matter having mass m. If such an event, together with the Earth chunk's having mass M, formed a complete cause of the gravitational force's occurring on Earth, then, weirdly, spatial locality would apply: the causes and effect would all be at the same place. But we can rule out this cheap way of satisfying spatial locality by noticing that in order for such an event to be a *cause* of the gravitational force on Earth, one of this event's *non-intrinsic* properties would have to be *causally relevant* to the gravitational force.

Let's take this more slowly. At the end of section 2, I mentioned that if C causes E, then only some of C's properties are "causally relevant" – that is, enable C to cause E. If C is lightning's striking the tree and E is fire's igniting, then C's being an event I am thinking about right now, one of C's *non-intrinsic* properties, is *not* causally relevant to E. Here's another example: If E is the window's shattering, then C's causally relevant features include its involving a hard object colliding with the window. C's occurring at the same instant as Jones's screaming "Not my window!" is one of C's *relations* to other events – one of C's *non-intrinsic* properties. It is *not* causally relevant to E. Likewise, only some of E's properties are causally relevant to C (that is, enable E to be caused by C). E's occurring exactly where C occurs is one of E's causally relevant, non-intrinsic properties.

Considering examples like these, I will assume that if C is a cause of E, then in connection with this causal relation, C's and E's causally relevant features must be some of C's *intrinsic* properties, some of E's *intrinsic* properties, and some of the spatiotemporal relations between C and E (such as their being separated by a distance r and occurring simultaneously) – and that is all! None of C's or E's *non-intrinsic* properties can be causally relevant, with the sole exception of C's and E's spatial and temporal relations to each other.

Suppose a complete cause of the gravitational force at t on a chunk of Earth were that chunk's possessing mass M at t and some other event C taking place *at that region of the Earth* at t. Then given Newton's law of gravity, C's causally relevant properties must include its occurring at a distance r from matter with mass m. But this is not an *intrinsic* property of C, since it has nothing to do with what is happening where and when C occurs (on Earth at t). No matter what is happening on Earth at t, it would be happening at a distance r from matter with mass m. Furthermore, C's occurring at a distance r from matter with mass m is not C's standing in some spatiotemporal relation *to E* (the gravitational force's occurring). So C, the event on Earth, cannot be a cause of E (Kim 1974; Lewis 1986: 262–4).

The same argument applies to the other example I gave earlier, in which a body passes through some location leaving no trace behind, but its having passed through there affects a second body passing through one hour later. Any event C occurring at the given location when the second body passes through automatically has the property of occurring one hour after the first body passed through. But this is not one of C's *intrinsic* properties (or C's spatiotemporal relation to the effect on the second body). So it cannot make C a cause of that effect. The events occurring at the given location when the second body passes through could (without any logical contradiction) have been *intrinsically* just as they actually were even without the first body's having passed through an hour before.

I have assumed that an effect's causally relevant properties must be intrinsic to it or involve its spatiotemporal relations to its cause. For example, suppose Jones lives in Seattle but his only brother lives in Chicago. One day, Jones's brother is killed in an automobile crash. Instantly, Jones becomes brotherless. If the crash *causes* Jones's suddenly being brotherless, then spatial locality has been violated: no chain of intermediate causes crosses the spatial gap between cause and effect. But this is not supposed to be a case of action at a distance. How can we keep spatial locality from being violated so easily?

We can do so by recognizing that Jones's being brotherless is not a property that Jones possesses *intrinsically*, since the property of having a brother is non-intrinsic (as I mentioned earlier). So Jones's becoming brotherless is not something that occurs at the time of the crash entirely in virtue of what is going on then *where Jones is*. Jones's becoming brotherless is not an *intrinsic* feature of any event occurring at Jones's location at the moment of the crash. Suppose Jones was eating a sandwich at that moment. Imagine him: there he is, becoming brotherless before your eyes. Did you see anything happen to him that was *intrinsically* his becoming brotherless? No! Everything happening at Jones's location could have happened just as it did (as far as its intrinsic properties are concerned) without Jones becoming brotherless. So Jones's becoming brotherless cannot be an event's causally relevant feature. But it would have to be, for the crash to cause Jones's becoming brotherless. So the crash does not *cause* Jones's becoming brotherless, though of course, the crash caused the death of Jones's brother. The event of Jones's turning brotherless consists of whatever is happening to Jones at that moment (say, his biting his sandwich, his heart beating, his moving at 3 m/s). That event is not intrinsically a becoming brotherless; it has causes, but among its causally relevant features is not its involving Jones's turning brotherless.

Thus, spatial locality cannot be undermined so easily. When a 5 kg body moves to a location 5 m away, I instantly acquire the property of being 5 m away from a 5 kg body. Is my acquiring this property *caused* by the body's moving to a certain location 5 m away? If so, spatial locality would be violated. But this is surely not a case of action at a distance! To see why, suppose for the sake of argument that this *is* a case of cause and effect, and then watch how this supposition leads to an impossible result. Under this supposition, the causally relevant features of the effect must be its involving my being 5 m away from a 5 kg body. But this is not an *intrinsic* property of anything happening where I am. Every event (E) happening here (such as my taking a breath) could have happened just as it did, as far as its intrinsic properties are concerned, without the 5 kg body being 5 m away (C). So for any event E happening where I am, *none* of its intrinsic properties is causally relevant to C. But this, we said earlier, is impossible: if C is a cause of E, then C's and E's causally relevant features must include some of C's intrinsic properties and some of E's intrinsic properties, as well as some of C's and E's spatiotemporal relations to each other. For if C causes E, then some of E's intrinsic properties must tie E to C. It cannot be that C would have caused E no matter what E was like intrinsically!

Strictly speaking, nothing *causes* me to acquire the property of being 5 m away from a 5 kg body. But the property's instantiation is not spooky or inexplicable. The explanation is humdrum: that a certain 5 kg body moved to a certain location, while I occupied a certain other location, explains why I acquired the property of being 5 m away from a 5 kg body. If you insist that the 5 kg body's arriving 5 m away from me *causes* me to be 5 m away from a 5 kg body, then I ask you: via what fundamental force – gravity, electromagnetism, the "strong" and "weak" nuclear forces – does the distant body affect me?

The Sun's causing gravitational forces on Earth stands in violation of spatial locality unless the forces have a local cause: the Sun's gravitational field on Earth. In the next chapter, we will begin to look more carefully at fields.[8]

Discussion Questions

You might think about . . .

1 Can you think of some examples – perhaps actual, perhaps imaginary, but in some way different from those given above – in which spatial locality holds but temporal locality is violated? What about where temporal locality holds but spatial locality is violated? Extra credit for cases where spatial locality *and* temporal locality hold but spatiotemporal locality is violated.

2 *In Sartor Resartus*, Thomas Carlyle (Scottish essayist, 1795–1881) wrote:

> It is a mathematical fact that the casting of this pebble from my hand alters the center of gravity of the universe.

Is this a violation of spatial locality? (Interpret "universe" as "solar system.")

3 Consider:

> [A]ny continuous medium . . . , if it consists of material particles, cannot provide an ontological basis for an alternative mode of transmitting actions to the mode of action at a distance. The reason . . . is that every particle with a sharp boundary always acts [in a collision] upon something outside itself, and thus acts where it is not.   (Cao 1997: 28)

Is this "acting where it is not" a violation of "spatial locality" as defined in this chapter?

4   Return to the single billiard ball in empty space feeling no forces from $t = 0$ to $t = 5$. Its trajectory from $t > 0$ to $t = 5$ is caused by its feeling no forces during that interval along with its position and velocity at $t = 0$. Its velocity at $t = 0$ is defined as the speed and direction of its motion at that instant. Since speed = (distance covered)/(time taken in covering it), a body's speed at $t = 0$ is usually defined as the "limit," as t approaches 0, of $[r(t) - r(0)]/t$. In other words, take the body's position $r(t)$ at some time t and its position $r(0)$ at time 0. Take the distance between them (which is how far the body moves in a period lasting for t), divide this by t (how long that period lasts), and see what value this quotient converges upon as the period around $t = 0$ shortens. So the body's speed at time 0 is nothing over and above a certain mathematical function of the body's positions at various moments surrounding time 0. Some of these moments occur *after* time 0. So the body's speed at time 0 is *constituted* partly by the body's having various positions at various moments shortly after time 0. But the body's having those positions at those moments are among the events that the body's speed at time 0 was supposed to help cause. The body's having those positions at those moments cannot be one of its own causes! How can we avoid this result?

Notes

1   This broad sense of "cause" is common in philosophy (Lewis 1986: 162). Conversational context influences what counts as *the* cause of (say) a police officer's death: the gunshot, failure to wear a bulletproof vest, or blood loss. Each is "causally relevant."

2   The broad sense of "event" is common in philosophy (Lewis 1986: 216).

3   Similar remarks apply if a fact is a state of affairs – a situation the obtaining of which *makes* a sentence true.

4   See Kline and Matheson (1987), from which I stole my next remark. Another puzzle concerning "action by contact" that I'm not going to explore is that a collision between perfectly hard bodies would result in an instantaneous, finite change in velocity – an infinite acceleration.

5   The body's moving at 5 m/s at $t = 2.5$, and the body's feeling no force between $t = 2.5$ and $t = 3$ (including those boundary moments), presumably constitute a complete cause of the body's moving at 5 m/s at $t = 3$. Of course, we would ordinarily not refer to the body's feeling *no* force as a "cause." We might prefer to say that a body continues to move at a constant speed and in a constant direction because *nothing* causes it to accelerate.

But the body's feeling no force is surely causally relevant, just like the tree's being dry and surrounded by oxygen. What is intuitively a *lack* (the body's feeling no force) *can* be a cause, as when a cause of the barn's burning down was the lack of a prompt response by the fire department.

6 Of course, we would expect a cause of E *not* to begin *after* E has already ended. But I see no reason to build this notion into the definition of "temporal locality." Some philosophers have seriously entertained the possibility of "backward causation" (Price 1996).

7 This is similar (but not identical) to Shoemaker (1969: 376).

8 We will not discuss a host of other interesting questions about causal relations. See Sosa and Tooley (1993).

2

Fields to the Rescue?

In this chapter, we will see that electric and magnetic interactions obey spatiotemporal locality *if* electric and magnetic fields are real things. Accordingly, our attention will turn to fields. We will explore some of the ways in which remarks about fields might be *interpreted* (as I mentioned in the Preface). In later chapters, we will examine arguments for and against interpreting fields as real – as existing on a par with matter. In the final sections of this chapter, we will examine two other kinds of entities (potentials and lines of force) the reality of either of which would make electric and magnetic interactions local. However, we will find fairly powerful reasons not to believe them real. The reality of fields will then be locality's best hope.

1 The Electric Force

In chapter 1, we saw that gravitational forces seem to be caused "at a distance" – across a gap in space. The same applies to electric and magnetic forces. One electrically charged body causes another, a distance r away, to feel a certain force. This force seems to have no cause located any nearer to it than r, violating spatial locality.

To discuss this properly, we must bear in mind some of the physics of electric forces, as would be found in any textbook of "classical" physics – that is, physics based on theories developed before the advent of relativity and quantum mechanics. A force is a *vector*: it has a magnitude (that is, an amount) and it pushes or pulls in a certain direction. Imagine two bodies the size of points, and suppose that for a long time, they have possessed electric charges q_1 and q_2, respectively, and been held at rest at a separation of r. Then *Coulomb's law* says that each body exerts on the other an electric force pointing away from the body exerting it and having a magnitude proportional to the product of the charges and inversely proportional to the square of the bodies' separation $(q_1 q_2 / r^2)$. (So if the bodies were twice as far apart, the force between them would be reduced to one-quarter of its actual value.) Electric charges can be positive (+) or negative (–). The product $q_1 q_2$ will be positive when the charges are of the same sign (producing a repulsive force) and negative when the charges are of different signs (producing an attractive force). The force that the first body exerts on the second is equal and opposite to the force that the second exerts on the first. (This accords with Newton's third law of motion: "action equals reaction.") Suppose we measure r in centimeters and the force in dynes. (One "dyne" in classical physics is the force whose constant application raises a 1 gram body's speed each second by 1 centimeter per second.) Then we can define a charge of 1 "statcoulomb" so that the force between two bodies, each charged to 1 statcoulomb and separated by 1 centimeter, is 1 dyne. In these units, Coulomb's law is

$$\boldsymbol{F} = [q_1 q_2 / r^2] \, \boldsymbol{u} \qquad\qquad (2.1)$$

where **bold** indicates a vector, $\boldsymbol{F}$ is the electric force on body #1 exerted by body #2, and $\boldsymbol{u}$ is a vector having a magnitude of one unit and pointing from body #1 directly away from body #2 (the direction in which #1 would be repelled by #2 were they charged alike).[1]

The electric force that body #2 exerts on body #1 is entirely independent of whether there are any other charges in the area. The acceleration $\boldsymbol{a}$ of #1 (the rate at which its velocity is changing) depends upon the *total* force that #1 feels (according to Newton's second law: $\boldsymbol{F} = m\boldsymbol{a}$, where $\boldsymbol{F}$ is the total force felt by the body and m is its mass). The total electric force on #1 is the vector sum of the electric forces exerted on #1 by all charged bodies (figure 2.1). According to Coulomb's law, even a point charge that is *very* far away from body #1 exerts a (perhaps tiny) electric force on #1. So spatial locality is apparently violated: the gap between cause and effect can be arbitrarily wide.

Coulomb's law is restricted to *point* bodies (see box 2.1). The total electric force exerted by one *extended* charged body on the other is the sum of the various forces between all of the point-sized "regions" comprising the two bodies. Coulomb's law is also restricted to *static* cases: a

Box 2.1
Coulomb's law generalized to extended bodies

Suppose the elementary charged bodies are not point-sized but have finite volumes. Imagine dividing each body into small non-overlapping regions. Take the charge $q(v_1)$ in a given volume v_1 within body #1 and the charge $q(v_2)$ in a given volume v_2 within body #2. If r_{12} is the distance between the centers of the two regions, then the vector quantity

$$[q(v_1)\ q(v_2)/r_{12}^2]\ \boldsymbol{u}$$

represents the contribution of v_1 and v_2 to the electric force between the bodies. Add all of these contributions, for every pair of regions from the two bodies, and then consider the limit to which this sum approaches as the bodies are divided into smaller and smaller regions. This is the electric force between the two bodies.

Instead of taking the sum first and then the limit, the same generalization of Coulomb's law can be expressed by taking the limit first and then the sum. This form uses the "charge density" $\rho(r)$ as a function of position r. Take a small region around a given point in an extended charged body. Consider the total charge in that region divided by the region's volume. Take the limit approached by that quotient as the region becomes smaller and smaller, contracting around the given point. The quotient's limit is ρ at that point. (Charge density comes in units of statcoulombs per cubic centimeter.) If a certain region has uniform charge density, then the total charge in the region is the region's volume times that charge density. Accordingly, a "region" v of infinitesimal volume dV contains total charge $\rho(v)\ dV$. The force between bodies #1 and #2 is the sum of all of the vector quantities

$$[\rho(v_1)\ dV_1\ \rho(v_2)\ dV_2/r_{12}^2]\ \boldsymbol{u},$$

one for each pair of infinitesimal volumes from the two bodies. Calculus expresses this sum as the double integral

$$\iint [\rho(v_1)\ \rho(v_2)/r_{12}^2]\ \boldsymbol{u}\ dV_1\ dV_2$$

taken over the volumes of both bodies.

Even if there are no point charges, they may remain a useful approximation and idealization.

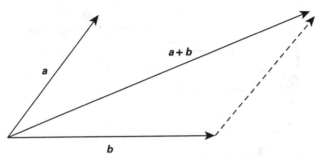

Figure 2.1 To add vectors **a** and **b**, move **a** so that it starts from the tip of **b**'s arrow. (The moved **a** is shown as a dotted line.) Form a vector starting where **b** starts and ending at the tip of the moved **a**'s arrow. This is **a** + **b**.

pair of point bodies that have possessed their charges and remained stationary for a long time. This restriction arises from the fact that the electric force felt now by body #1 depends not on body #2's charge and separation from #1 *now*, but on #2's charge *sometime earlier* and the distance between #1's location now and #2's location *then*. (In a static case, we can ignore this time delay since the earlier values equal the current ones.) In particular, if the bodies are held at rest, separated by a distance r, and #2's charge somehow doubles, its force on #1 increases, but not instantly – only after a delay of r/c seconds, where c is the speed of light in a vacuum (about 300,000,000 meters per second).

You might be tempted to say that the "news" of body #2's increased charge takes some time to reach body #1. The "news" travels at the speed of light, so to cross the bodies' separation r, the "news" requires r/c seconds. If the news's arriving at #1's location at t causes #1's behavior at t, then spatiotemporal locality holds. However, we have not been imagining that some actual *thing* moves from #2 to #1 across the gap between them. Rather, we have been thinking of electric forces as operating by action at a distance: no cause of #2's force on #1 lies less than r from #1. Between the cause and its effect, there is a gap not only in space but also in time, lasting r/c seconds, since #2's force on #1 at t has no complete set of causes occurring within r/c seconds of t. We have here not *instantaneous* action at a distance, but rather so-called *retarded* (that is, time-delayed) action at a distance.

It was known by about the mid-nineteenth century that any action at a distance involving electric or magnetic forces would be *retarded* and so undermine not only spatial locality, but also temporal locality. In contrast, classical physics never determined gravitational forces to be

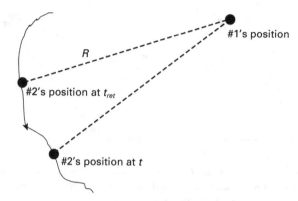

Figure 2.2   The path of body #2 as it passes near body #1, showing #2's position and "retarded position" at time *t*.

non-instantaneous.[2] Therefore, in examining locality, I have turned my attention to electromagnetic forces.

Let's generalize Coulomb's law to cover the electric force at *t* on body #1 (which has long possessed charge q_1 and been at rest) exerted by body #2, which may have been in any state of motion whatever (but has long possessed charge q_2). There was some earlier moment t_{ret} such that "news" of #2's condition at t_{ret} is the "news" just reaching #1 at *t* (figure 2.2). Body #2's position at t_{ret} is its "retarded position" at *t*. Let *R* be the distance between #1's position at *t* and #2's retarded position at *t*. Then $R = c(t - t_{ret})$, so "news" from t_{ret}, traveling at speed *c*, completes its journey across separation *R* exactly at *t*. (Again, this talk of "news" is misleading insofar as it suggests local action rather than action at a distance.) The general law for the electric force that #1 feels at *t* as a result of #2 is a somewhat complicated expression (Feynman et al. 1963: II, p.21-1):

$$\boldsymbol{F} = q_1 q_2 [\boldsymbol{u}/R^2 + (R/c)\,(\boldsymbol{u}/R^2)' + \boldsymbol{u}''/c^2] \tag{2.2}$$

where $\boldsymbol{u}$ is the unit vector pointing from #1 directly away from #2's retarded position. A prime (') following a symbol means the symbolized quantity's rate of change; a double-prime (") means the rate of change of that rate of change. The first term – $q_1 q_2\,(\boldsymbol{u}/R^2)$ – is just Coulomb's law for the retarded position. In the static case, the retarded position is just the current position (and all of the primed and double-primed terms equal zero), so we get Coulomb's law back again.

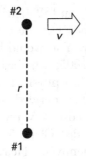

Figure 2.3 Body #2 has been moving to the right at speed v and now passes body #1 at a distance r.

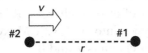

Figure 2.4 Body #2 has been moving to the right at speed v and now reaches a distance r from body #1.

The second term is the time delay (R/c) multiplied by the rate of change of the retarded Coulomb force. The third term involves #2's acceleration. The important point is that the force on #1 at t is caused by #2's charge, position, and motion at the retarded time t_{ret}.

Here is an application of this formula that is easy to picture. Imagine that for a long time, body #1 has been at rest whereas #2 has been moving uniformly – that is, in a constant direction at a constant speed v (less than c). When #2 streaks past #1 at a distance r (as shown in figure 2.3), #2's electric force on #1 equals

$$[q_1 q_2/r^2]/\sqrt{[1 - (v^2/c^2)]} \tag{2.3}$$

directed away from #2. Since $\sqrt{[1 - (v^2/c^2)]}$ is less than 1, the force is greater than $[q_1 q_2/r^2]$, the value given by Coulomb's law. As arranged in figure 2.4, #2's electric force on #1 equals

$$[q_1 q_2/r^2] \, [1 - (v^2/c^2)] \tag{2.4}$$

directed away from #2, which is less than the value given by Coulomb's law. (Remarkably, the force on #1 turns out to be directed away from #2's *current* position even though the force is *retarded*.)

Examining equations (2.3) and (2.4) with their $[1 - (v^2/c^2)]$s, you may be thinking that they look like the sorts of expressions found in Einstein's theory of relativity. You would be right. Yet relativity was formulated by Einstein in 1905, whereas these formulae were discovered in 1888 by the English physicist Oliver Heaviside (1925: II, 494–6). Heaviside (1850–1925) derived them from the basic equations of classical electromagnetism, "Maxwell's equations," originated by the Scottish physicist James Clerk Maxwell (1831–79) and put into modern form independently by Heaviside and Heinrich Hertz (German physicist, 1857–94). The point is that Maxwell's electromagnetic theory was the single piece of classical physics that was already relativistic before Einstein. It had to be, since magnetic forces turn out to be a relativistic effect of electric charges. Relativity theory brought the *rest* of classical physics into line with Maxwell's electromagnetic theory, as we will see in chapters 7 and 8.

2   The Electric Field and its Possible Interpretations

According to equation (2.2), body #2's electric force on body #1 can be computed by multiplying #1's charge by a complicated factor involving various quantities pertaining only to #2. This factor may differ from place to place at a given moment, and from moment to moment at a given place. This factor (a vector) is #2's "electric field" *E*. In other words, #2's electric field at a given location and moment equals the electric force that #2 would exert on a point body at that place and moment for each unit of that point body's electric charge:

$$\boldsymbol{F} = q\boldsymbol{E}. \qquad (2.5)$$

(The field's strength E is some number of dynes per statcoulomb.) Whereas #2 causes an electric *force* only at locations where there exist charged bodies, #2's electric *field* can have a non-zero magnitude even in "empty space" – that is to say, empty of matter.

Here's an example. Consider a point body with charge q that has been stationary forever. Then at a given moment at some location (a distance r away from the body), the body's field *E* is equal (in dynes per statcoulomb) to the electric force (in dynes) that the body would exert on a 1-statcoulomb point body there then, as dictated by Coulomb's law. So

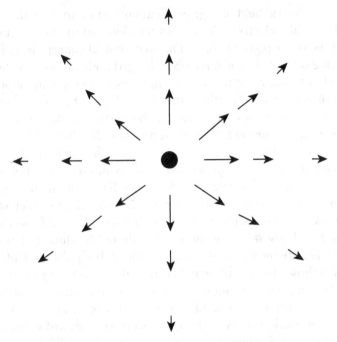

Figure 2.5 A typical textbook representation of the electric field of a positive charge (the dot), crudely displaying its direction and diminution with distance.

$$\boldsymbol{E} = [q/r^2]\,\boldsymbol{u} \tag{2.6}$$

where $\boldsymbol{u}$ is a unit vector from the given location pointing away from the charge whose field it is. (See figure 2.5 for a depiction of the "Coulomb field.")

Now suppose that the point charge q has not been stationary forever. Rather, it was stationary until a certain time t_1, when it moved (during a negligibly short interval of time) to another location, from which it has not strayed since. In the time interval from t_1 to now (time t), news of its new position will have spread (at speed c) in all directions from the new location out to a distance $c(t - t_1)$. This news is embodied in the field. The field surrounding the point charge's current location, out to a distance $c(t - t_1)$, is the Coulomb field. There is also a Coulomb field centered on the point charge's earlier location, except that this field is zero out to a radius of $c(t - t_1)$ around that location. Beyond that distance, news has not yet arrived that the body has moved from its earlier location. The field at a given moment is a standing record of the arrangements of charges in the past.

The net electric field at a given location and moment is the vector sum of all of the electric fields at that location and moment caused by each of the other point charges. The electric-field concept is useful for computing the net electric force on a charged body because the field at that body's location captures in a single vector everything about the surrounding charges that affects that force there. Once you know the field, you no longer need to consider the particular configurations of charges, past and present, that are responsible for that field.

The electric field would enable electric interactions to satisfy spatiotemporal locality – *if* the field were real, an immaterial object existing on a par with matter. Fields and bodies would then constitute the "furniture of the universe" (though see box 2.2). In other words, we saw earlier that spatiotemporal locality would hold if the electric force at *t* on body #1 were caused directly not by distant charges at some earlier moment, but by the arrival at body #1 at *t* of news regarding those charges. If the field is real, then *it constitutes the news*: the field vector at a given location is a trace left by the behaviors of distant charges at earlier moments. Like a pebble dropped into a pond and causing concentric ripples in the water to travel outward to the edges of the pond, diminishing in intensity as they travel farther away, a point charge causes an event that moves outward from the charge in all directions (at speed *c* through empty space). This effect event consists of some contribution to the field. It's like a "ripple" in the field; the event moves, like the event of my car's slowing down to a stop. The effect event begins at the time of its cause (the point body's occupying a certain location, with a certain velocity, acceleration, and quantity of charge) and at the same place as its cause. So spatio-temporal locality is satisfied. The electric force on some point charge at *t* is caused by the field at its location at *t*, again satisfying locality.

If the electric field is real and other forces operate through other fields, then we can resolve a problem we encountered in chapter 1: how to understand collisions. We found it difficult to understand how two bodies could touch: being material, they cannot share a point, but if they do not occupy the same place at the same time, then they must be separated by at least one point, and so there is a gap between them even while they are supposedly in contact. Fields can rescue locality from this difficulty. Suppose each particle of matter is surrounded by a very short-range field of highly repulsive force. (Let's not worry about whether this force is electric or of some other kind.) The field surrounding a given particle then causes an immense repulsive force on

Box 2.2
Fields and the aether

The distinction between fields and matter was not drawn sharply when field theory originated. The field at a given location was then considered to be a condition there of the "aether," a stuff pervading all of space. An analogy may help: hydrodynamics, the study of fluid flow, describes the "velocity field" of a flowing liquid. This field's vector at a given location is the speed and direction of the motion of the liquid's particles there. Of course, there is no velocity field over and above the liquid's moving particles; remarks about the velocity field are really made true or false by the motions of the liquid's particles. Likewise, when the electric field was introduced, scientists considered remarks about the electric field to be about the state of the aether, a material entity.

Sometimes, the aether was thought to consist of small particles near to but not touching one another. Then the "existence" of fields (as conditions of the aether) would not eliminate action at a distance from electrical interactions. One charged bit of ordinary matter would affect the aether particles nearest it, and each of them would in turn affect the next aether particles, and so on. In this way, the effect would eventually reach distant charged bodies. But since the aether particles do not touch one another, each step in the process would constitute action at a (small) distance, violating spatiotemporal locality (Maxwell 1890: II, 485–6). On the other hand, spatial locality could be upheld by an aether that was *continuous* rather than made of particles.

As we will see, physics textbooks today ordinarily portray fields as real entities, possessing their own properties at various locations, rather than as properties of something else (such as the aether). The reality of such fields would eliminate action at a distance.

any particle that comes very near to the given particle. When two particles "collide," they do not touch, but they come near enough for each to feel a tremendous repulsive force from the other's field. So the two particles appear to bounce off each other. Each interacts with the field *at that moment at its own location*, so locality is upheld. Although one body excludes any other from the locations it occupies, those locations *can* be occupied simultaneously by a field and matter.

Other issues may be harder to deal with. Equation (2.6) gives the field of a stationary point charge. The field's magnitude becomes infinite when $r = 0$ (since we are then dividing by zero), and the direction *u* becomes indeterminate at the point charge's own location (since there, *every* direction points away from the point charge, just as at the north pole, every direction on Earth is south). If the field is a real thing, then presumably it cannot become infinite in strength and indeterminate in direction at some location. Furthermore, if the point charge contributes infinitely to the field at its own location, then that contribution would drown out the contributions of other, distant charges. The point charge would then never be attracted or repelled by other bodies! So the contribution that a point charge makes to *E* at $r = 0$ must be zero. But this must be inserted "by hand"; Maxwell's fundamental equations of electromagnetism do not tell us this (Feynman et al. 1963: I, p.28-3). Of course, if fields are just mathematical devices we use for predicting the electric forces on bodies, then we can just insert this result by making it one of the rules for using the device. But if fields are real, this seems artificial: why does a body's field, after growing arbitrarily great as r becomes arbitrarily close to zero, suddenly vanish utterly at $r = 0$?

If there are no *point* charges (that is, if the smallest charged body has finite non-zero volume), then this problem fails to arise. But another still does if fields are real. Does each fundamental charged body have its own electric field, overlapping with the electric fields of other fundamental charged bodies? Or is there a single electric field, extending throughout space, to which all fundamental charged bodies contribute? If the electric field is more than a calculational aid, then there must be a definite fact about whether there is one electric field or many. If there are many, then we could solve the $r = 0$ problem for point charges by holding that a charged point particle is acted upon only by other bodies' fields (so-called "incident" or "external" fields), not by its own. But if there is only one field, then that field threatens to become infinite at any point charge's location.

We could avoid having to face the one-field-or-many question by not thinking of electric fields as real things. The field may be a useful fiction, just a convenient device for figuring out the electric force on a body. (Then spatiotemporal locality would again be violated; electric interactions would involve retarded action at a distance.) The remark "*E* here now is 5 dynes/statcoulomb" would then be false if interpreted *literally* – as aiming to describe a certain real entity here: the electric field. The remark might still be very useful to make, and it might even be true if interpreted *non-literally*. For instance, it might be interpreted as saying that a point charge, *were* one here now, *would* feel 5 dynes of electric force for each statcoulomb of its charge. For this to be true, there need not actually be a point charge here now. But if there happens to be one, the remark about E does not specify a local *cause* of the force it feels. (I will investigate why in chapter 3.)

Under a different non-literal interpretation, "E here now is 5 dynes/ statcoulomb" means that 5 dynes per statcoulomb is what results from a certain combination of multiplications, divisions, and additions of various quantities concerning distant charged bodies at retarded times. For instance, suppose we are talking about the field of a charged point body that has been stationary for a long time, so equation (2.6) applies. Then under this non-literal interpretation, the remark "E here now is 5 dynes/statcoulomb" means that 5 dynes per statcoulomb is the result of dividing the body's charge by the square of the distance to the body. So interpreted, the remark may be true – not because of the state of some immaterial thing, the electric field, but simply in virtue of the body's charge and distance. A remark about the electric field is then shorthand for what may be a fairly involved arithmetical expression, considering the complexity of equation (2.2). A similar interpretation should be given to the remark "The average family in the USA has 2.1 children." It describes the result of a certain arithmetical operation: dividing the total number of children living in families in the USA by the total number of families in the USA. This remark is made true not by a certain real family in the USA (the "average" one) existing alongside all of the "ordinary" families in the USA, but rather by certain facts about those "ordinary" families. On this interpretation of the electric field, the remark "E here now is 5 dynes per statcoulomb" again does not specify a *local* cause of the force felt by a charged body here now. Instead, this remark refers to various events involving distant bodies at retarded times.

The magnetic field B is similar in many respects to the electric field E, and we have the same options for interpreting remarks about it. The magnetic field is often illustrated by arrays of iron filings, each filing aligned with B at its location (figure P.1). Of course, a location occupied by an iron filing is not *empty* space. But it is easy to imagine that the iron filing is responding to a condition of the space it occupies that would be present whether or not an iron filing were there to respond to it. In other words, it is difficult not to assume spatiotemporal locality. Whereas the electric force on a point charge q is qE (equation 2.5), the magnetic force F depends on the point charge's velocity v,

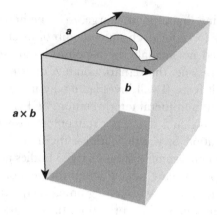

Figure 2.6 The length of the vector **a × b** (the "cross product" of **a** with **b**) is **a**'s length times **b**'s length times the sine of the angle between them. (An angle's sine falls somewhere between 1 and −1. Since the sine of 90° is 1, **a × b** is greatest when **a** is perpendicular to **b**, and since the sine of 0° is zero, **a × b** is zero when **a** is parallel to **b**.) The direction of the vector **a × b** is perpendicular to both **a** and **b**. If we must turn **a** clockwise (as shown in the diagram) to turn it into **b** in the shortest way, then the direction of **a × b** is *into* the imaginary clock having **a** and **b** as its hands. If we must turn **a** counterclockwise, then the direction of **a × b** is *out* of the clock. The magnetic force on a moving particle, then, is perpendicular to both its velocity and the magnetic field **B** at its location.

disappearing entirely when the body is stationary. (The magnitude of a body's velocity vector **v** is the body's speed; **v**'s direction is the direction in which the body is moving.) Specifically:

$$\mathbf{F} = (q/c)\mathbf{v} \times \mathbf{B}. \qquad (2.7)$$

(The × here represents a kind of multiplication: the "cross product" of **v** with **B**. This is explained in figure 2.6.)

Let's summarize. We are dealing with two questions: "Does spatiotemporal locality hold?" and "Are fields real?" To say that fields are real is to interpret the remark "*E* here now is 5 dynes per statcoulomb" *literally*: as describing a property possessed by a certain real thing, the electric field. On the other hand, to say that fields are unreal is to interpret the remark *non-literally*. For example, the remark might be interpreted as describing the electric force that would be felt by a charged body, were a charged body present. Alternatively, it might be interpreted as describing something about the arrangement of distant charges at earlier moments. These three interpretations disagree regarding what state of affairs *makes true* the remark "*E* here now is 5 dynes per statcoulomb." On the first interpretation, the remark is true by virtue of a property possessed here now by the electric field. On the second interpretation, the remark is made true by a fact about what would have happened had certain events occurred that may not, in fact, have occurred. On the third interpretation, the remark is made true by certain events distant in space and time. If electric and magnetic fields are real, then spatiotemporal locality applies to electromagnetic interactions: the force on a charged body now is caused by the field's condition now at the body's location. (Fields act on charged bodies but *not* on one another.) But if fields are convenient fictions – in other words: "auxiliaries," practical aids to calculation, mere theoretical devices, mathematical "artifacts," formal tools lacking "physical meaning" – then they are powerless to oppose the thought that the effect of one charged body (or magnet) on a distant charged body constitutes retarded action at a distance, violating both spatial and temporal locality. So which is it – are fields real or not?[3]

In physics textbooks, remarks asserting the field's reality are common. Here is a typical excerpt, taken from one of my own college texts:

> We think of the electric field as a condition in space set up by the system of point charges. This condition is described by the vector $\boldsymbol{E}$. By moving the test charge q_0 from point to point we can [by measuring the electric force on the test charge and using equation (2.5)] find the electric field vector $\boldsymbol{E}$ at any point (except one occupied by a charge q_i). . . . [T]he electric field is more than a calculational device. This concept enables us to avoid the problem of action at a distance. . . . We thus think of the force exerted on a charge q_0 at point P as being exerted *by the field at point P* rather than by the charges, which are some distance away. Of course the field at point P is produced by the other charges, but not instantaneously.   (Tipler 1976: 705–6)

This sounds like an *argument* for the field's reality from the fact that it "enables us to avoid the *problem* of action at a distance." But we cannot simply *declare* action at a distance to be a "problem." Of course, action at a distance would be a problem *for spatiotemporal locality*. But why, precisely, should we take it for granted that spatiotemporal locality holds *generally* and not merely of billiard-ball collisions and other familiar interactions? In chapters 4 and 5, we will examine some possible *reasons* for calling action at a distance a "problem."

In presenting fields as real things, physics textbooks accurately reflect the attitudes of many physicists past and present, beginning with Michael Faraday (English physicist, 1791–1867) and Maxwell. Here, for instance, is a recollection of Maxwell's lectures in 1878:

> He was very fond of using words like "*the quality or peculiarity in virtue of which*".... So in defining a magnetic field as a region possessing or possessed of a peculiarity he cautioned us against thinking that a field of the strength of so many [units, namely, dynes per statcoulomb] meant so many dynes [that is, so many units of *force*]; the field is "there", whether or not an isolated unit magnetic pole be set in it to experience the force of so many dynes.   (Newall 1910: 106–7)

An iron filing's alignment to the surrounding magnetic field shows only that the filing feels a force. This force would not have existed had no matter been there. A force is always a force *on something*; there are no free-floating forces. According to Maxwell, however, the magnetic field exists at the filing's location and *would* still have existed there even if no body had been there for the field to push around. But if the magnetic field cannot be detected on its own and functions (according to Maxwell's theory) only as a causal intermediary between the magnet and the iron filings, then why not interpret the magnet as influencing the filings *directly* (by retarded action at a distance), with no "middle man"? What *reason* have we for thinking that *anything* exists over and above the bodies?

We will see some possible reasons in later chapters. For now, my point is simply to acquaint you with the standard view, according to which (in the words of a classic textbook)

> Maxwell's theory then goes on to ascribe to this vector E a self-existent reality independent of the presence of a testing body [that is, a body to be pushed around by the field]. Although no observable force appears unless at least two charged bodies are present ..., nevertheless we assert with Maxwell that the charged piece of metal by itself produces in the surrounding space a change of physical conditions which the field of the vector E is exactly fitted to describe. The primary cause of the action on a testing body is considered to be just this vector field *at the place where the testing body is situated*. As for the piece of metal, its part in the matter is merely to maintain this field. We speak accordingly of a theory of *field action*, as contrasted with the theory of *action at a distance*.   (Abraham 1951: 55)

Should we believe this? Some physics textbooks dissent:

> Any statement which is made about the electric field in the neighbour-hood of a charged body cannot strictly speaking be taken to mean more than that a second charged body, if placed there, would behave in a particular way. (Pilley 1933: 84)

This is one of the non-literal interpretations I mentioned.

Sometimes one finds remarks like these, by the American physicist and philosopher of science Percy Bridgman (1882–1961), 1946 Nobel laureate in physics for his experiments on high-pressure phenomena:

> The reality of the field is self-consciously inculcated in our elementary teaching, often with considerable difficulty for the student. This view is usually credited to Faraday and is considered the most fundamental concept of all modern electrical theory. Yet in spite of this I believe that a critical examination will show that the ascription of physical reality to the electric field is entirely without justification. I cannot find a single physical phenomenon or a single physical operation by which evidence of the existence of the field may be obtained independent of the operations which entered into the definition. The only physical evidence we ever have of the existence of a field is obtained by going there with an electric charge and observing the action on the charge . . . , which is precisely the operation of the definition. . . . The electromagnetic field itself is an invention and is never subject to direct observation. What we observe are material bodies with or without charges (including eventually in this category electrons), their positions, motions, and the forces to which they are subject. (1958: 57–8, 136)

Bridgman here treats the hypothesis that fields are real as a scientific theory. He says that this theory can tested by our observations, and so far, it has not been justified by them.[4] In Bridgman's view, this theory is similar to the theory that intelligent life exists elsewhere in the universe: there is, as yet, no observational evidence in its favor, since every observation so far can best be explained without appealing to a real field (or intelligent extraterrestrials).

According to another view that is sometimes encountered, the hypothesis that fields are real is (contrary to Bridgman) *not* the sort of hypothesis that science concerns itself with – not the sort of theory that could, even in principle, be tested by our observations. You *couldn't* have any evidence for the reality of fields independent of their effects on test charges, and those effects could just as well be explained by retarded action at a distance. Here's a statement of such a view from another physics textbook:

> The assertion [of the field's reality], taken by itself apart from the quantitative force-law, is scientifically otiose. [What a good word! It means "useless, doing no work."] It is merely the physically irrelevant statement of a metaphysical conviction. . . . This is certainly not a legitimate physical theory at all; it is the confusion of metaphysical belief with metrical physics. . . . The "field" may act as a metaphysical background, but it certainly does not act as a scientifically verifiable physical intermediary. . . . The cause [of the electric force on some body] may be all kinds of things, some of them rather queer; but we do not need to consider how it is brought about; in fact we have not got the faintest notion. The important point is that another charge if placed at that point would be acted upon by a force. It is not merely the important point; so far as physics is concerned, it is the only point.   (O'Railly 1965: II, 653–4)

Of course, you need a good *reason* for believing that some theory falls outside the bounds of science, just as you need a good reason for believing some theory true. Otherwise, your belief is unjustified; you hold it irresponsibly. We will have to see whether the field's reality would be "physically irrelevant."

3   Potentials

Besides fields, two other entities are sometimes mentioned in physics textbooks as if they existed in space empty of matter, allowing a magnet or electrically charged body to exert a force on distant charged bodies. These entities are potentials and lines of force (sometimes called "field lines"). Let us consider whether they, rather than or in addition to fields, might be real things, not mere notational devices. If they were real, they would enable spatiotemporal locality to hold for electromagnetic interactions.

The simplest potential is the electric potential in a "static" case (that is, where any electric charges in motion have a negligible effect on the overall electric field). Whereas the electric *field* is a *vector* function of position, the electric *potential* is a *scalar* function of position: it has a magnitude but no direction. At a given location, the electric field vector points in

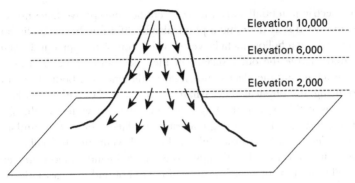

Elevation 10,000

Elevation 6,000

Elevation 2,000

Figure 2.7 A steep, fairly symmetric mountain, showing at various points the vector in the direction of steepest descent whose length reflects the steepness there. These vectors are analogous to the electric field vectors while the elevations are analogous to the values of the electric potential.

the direction in which the electric potential diminishes most steeply. To imagine this, picture some hilly terrain, with valleys, mountains, and so on (figure 2.7). The elevation at each point is like the value of the potential there. At each point, there is a direction that a ball would roll if put there: the direction of steepest downward slope. This is the electric field's direction there, that is, the direction of the electric force that a positive point charge there would feel. The magnitude of the potential's slope in that direction is the field's magnitude there.

For example, consider a stationary point charge q all by itself. At a distance r from the charge, the electric potential is q/r. Obviously, the potential decreases most steeply in the direction of increasing r – that is, in the direction pointing away from the charge. The rate at which the potential decreases in that direction is q/r^2. (For you calculus fans, that's the potential's derivative with respect to r, multiplied by −1 since we want the rate at which the potential *decreases*.) Thus we have arrived at the electric field around a stationary point charge (equation 2.6).

The electric potential is especially useful to discuss in connection with electric currents. Electrons (being negatively charged) flow through conductors from regions of lower potential to regions of higher potential. Such a potential difference is ordinarily called a difference in "voltage."

Although the electric field everywhere is determined by the electric potential everywhere, the potential in classical electromagnetic theory is typically regarded as merely a device for representing the field. In Maxwell's words:

> The electric potential ... is a mere scientific concept; we have no rea-
> son to regard it as denoting a physical state. If a number of bodies are
> placed within a hollow metallic vessel which completely surrounds them,
> we may charge the outer surface of the vessel and discharge it as we
> please [changing the electric potential of the enclosed bodies] without
> producing any physical effect whatever on the bodies within. ... Fara-
> day proved this by constructing a hollow cube, twelve feet in the side,
> covered with good conducting materials, insulated from the ground and
> highly electrified by a powerful machine. "I went into this cube," he
> says, "and lived in it, but though I used lighted candles, electrometers,
> and all other tests of electrical states, I could not find the least influence
> upon them."   (1881: 53; also 1995: 965–6; and Cat 1998)

Inside the cube, everything is at the *same* potential. Consequently, as
far as the events there are concerned, it makes no difference whether
the cube's potential is 200 volts higher than, or 100 volts lower than,
that of the ground.

The reason for interpreting the electric potential as a useful fiction is
ultimately that *many* electric potential functions correspond to the *same*
electric field. To see why, bear in mind that potential *differences* (the
potential's *slope*, its *rate of change*) determine the field. So the potential's
absolute values at various locations make no difference to the force that
a charged body would feel; only the *relative* values matter. If the potential's values were everywhere 5 volts higher, the field would be no
different than before, since the *differences* in potential between various
locations would be no different. Only the *difference* in potential between
a power line and the ground determines whether you will receive an
electric shock if you touch the line (that is, whether a current will flow
from the line through you to the ground). It does not matter whether
the electric potential at the line is 200 volts and at the ground is 300
volts or whether the line's potential is −100 volts and the ground's
potential is 0 volts. So there is no correct answer to the question: "Is
the power line's potential really 200 volts or −100 volts?"

Again picture the potential as elevation. The speed and direction of
a ball placed at a given location does not depend on the location's
elevation, but only on the terrain's slope there (the rate of change in
elevation). A location might be high but flat, in which case, the ball
would stay there. Or the location might be low but steep, in which
case, the ball would roll away. We usually think of elevation as relative
to sea level; when we say that Mount Everest's elevation is 29,028
feet, we mean "above the sea." But we could just as well have taken

any other point instead of the sea to be our zero level. For instance, elevation could have been expressed relative to the lowest point on Earth's landmasses: the Dead Sea. Since the Dead Sea is 1,286 feet below sea level, Mount Everest's summit is 30,314 feet above the Dead Sea. To ask whether Mount Everest's elevation is really 29,028 feet or 30,314 feet is to make a mistake: there is no fact about its *real* elevation. We can make no factual error by setting the zero level at one location rather than another. There are no *absolute* elevations; a location's elevation is not one of its *intrinsic* properties. There are facts only about elevation relative to a given reference point.

The same idea is thought to apply to the electric potential:

> Potential, in fact, is a *relative* term, the value of which can be expressed only by arbitrary definition, just as terrestrial longitude can be expressed only with reference to some arbitrarily-chosen meridian, such as that of Greenwich or Paris.   (Maxwell 1879: 271)

A location's potential (at a given moment) is thus not one of its intrinsic properties, and so (we saw in chapter 1) cannot be causally relevant to an electric force there. Of course, the *difference* in elevation between (say) Mount Everest and the Dead Sea is the same no matter what we take the zero point of elevation to be. (Relative to the level of the sea, the difference is $29,028 - (-1,286) = 30,314$. Relative to the Dead Sea, the difference is $30,314 - 0 = 30,314$.) Likewise, there is a fact regarding the potential *difference* between two locations. But what *makes* this fact hold is *not* the difference between the potential's absolute value at one location and its value at the other; there are no such values. So what state of affairs *makes true* the remark "The potential difference between these locations is 100 volts?" As we have seen, the electric field does not determine the potential's *absolute value* at a given location, but it determines the potential *difference* between two locations. So that difference holds in virtue of whatever state of affairs makes true some remark about **E**. This state of affairs involves the electric field itself, if the field is a real thing, or otherwise perhaps involves distant charges at earlier times.

Here is another version of this argument against the potential's reality. According to Maxwell's equations (as confirmed by Faraday's experiment), nothing depends on the electric potential's absolute value at a given location. Why is this? Because there is no such thing as the electric potential's absolute value at a given location! That would seem to best explain why nothing else would have been different if every location's electric potential had been 5 volts higher. Eugene Wigner (Hungarian-American, 1902–95), 1963 Nobel laureate in physics, draws a provocative analogy:

> [T]he potentials are not uniquely determined by the field; several potentials . . . give the same field. It follows that the potentials cannot be measurable, and, in fact, only such quantities can be measurable which are invariant under the transformations which are arbitrary in the potential. This invariance is, of course, an artificial one, similar to that which we could obtain by introducing into our equations the location of a ghost. The equations then must be invariant with respect to changes of the coordinate of that ghost. One does not see, in fact, what good the introduction of the coordinate of the ghost does.   (1967: 22)

Like the ghost's position, the electric potential's absolute value is not needed to account for anything. Isn't it reasonable, then, to conclude that the electric potential, like the ghost, is unreal?

This argument for the potential's unreality is fairly strong but certainly not utterly conclusive. That the potential's *absolute* values at various locations (over and above the potential's *relative* values) have no observable consequences in classical physics (as Faraday discovered in his cube) does not rule out the possibility that as physics progresses, we will discover some means of determining experimentally the potential's absolute value at a given location. As further discoveries are made, a quantity once thought observationally inaccessible may become detectable.

For example, temperature was once thought to be like potential and elevation in having an arbitrary zero point. (The zero on the Celsius scale is set at the freezing point of water in standard conditions but is set elsewhere on the Fahrenheit scale, relative to which water freezes at 32°.) Only temperature *differences* were thought to be real. Later discoveries revealed laws of nature connecting temperature to other concepts, thereby giving temperature a natural, absolute zero some 273 Celsius degrees below water's freezing point in standard conditions. Compare Faraday in his cube of uniform potential to someone in a container where everything has the same temperature, and so heat does not flow. Whereas Faraday observed nothing inside his cube that was affected by the potential's value there relative to the ground, someone inside the uniform temperature container (and aware of the relevant laws of nature) could tell something about its absolute temperature from, for instance, whether water there is liquid, ice, or steam.

But let us suppose, purely for the sake of argument, that Faraday's conclusion stands up – that there is no connection waiting to be discovered between the potential's absolute value and other facts about the world. The potential's absolute value would then have to be a kind of "dangler," a loose end, making no difference (through the laws of nature) to any other facts. Does this show conclusively that there is no fact of the matter regarding the potential's absolute value at a given location? Not unless it is impossible for there to be "danglers" of this kind. Why would that be? That the potential's absolute value would be a dangler shows only that the potential's absolute value is not something that we could ever know about – and (fortunately!) is not something that we would ever need to know about in order to predict the forces on bodies. It may be real, but we could never have a good reason for believing it real. (In subsequent chapters, we will see other arguments of this form and come to a better understanding of this argument's strengths and weaknesses.) Bertrand Russell (English mathematician and philosopher, 1872–1970) made a related point:

> [I]f anything occurs which is causally isolated, we cannot include it in physics. We have no ground whatever for saying that nothing is causally isolated, but we can never have ground for saying: Such-and-such a causally isolated event exists. The physical world is the world which is causally continuous with percepts, and what is not so continuous lies outside physics. (1927: 325)

Nevertheless, in chapter 8, we will encounter a quantity that for a long time was frequently considered real even while its absolute value appeared to be a "dangler": a body's absolute velocity.

4 Lines of Force

Let us turn to lines of force. A line of force is a curve with a directional arrow on it. The curve at a given location points in the direction of the field there – that is, in the direction that a positive point charge (in the case of electric lines of force – figure 2.8) or magnetic north pole (in the case of magnetic lines of force) would move, if placed there. A line of force bends smoothly. A magnetic line of force forms a closed curve or goes on forever, whereas an electric line of force begins at a positive charge and ends at a negative charge.

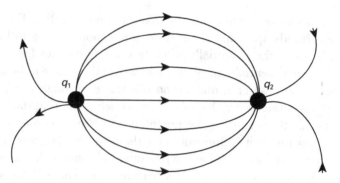

Figure 2.8 Electric lines of force between charges of equal magnitude: q_1 is positively charged, whereas q_2 is negatively charged.

Suppose that the line of force at a given spot at a given moment causes the force on a charged body there then, satisfying locality. The force's direction is determined by the direction of the line of force. What determines the force's strength, its intensity? The typical answer is that the density of the lines of force at any location – what Faraday calls "their concentration or separation, i.e., . . . their number in that space" (1852: 41) – is proportional to the force that a charged body there would feel. But in a region where the field is non-zero, each point lies on exactly one line of force. (If there were more than one electric line there, then the net electric force on a body there would lack a well-defined direction.) So the number of lines in a region where E is everywhere non-zero fails to increase as E increases there. A typical textbook diagram (figure 2.8) or depiction of iron filings (figure P.1) shows not *every* line of force, but only a representative sample: through some points where the field is non-zero, no line is shown.

The lines of force can be related to the intensity of the corresponding force by a tactic that Maxwell developed. Think of the lines as forming the surfaces of tubes through which an incompressible fluid flows (figure 2.9). No fluid crosses a tube's surface. (That the fluid is "incompressible" means that it has the same density wherever it is.) The fluid's speed and direction determine respectively the magnitude and direction of the field – and so of the force that would be felt by a unit positive electric charge (or unit north magnetic pole) at that location. This fluid, Maxwell (1890: I, 160) says, is "not even a hypothetical fluid which is introduced to explain actual phenomena. It

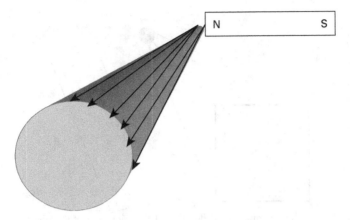

Figure 2.9 Some magnetic lines of force helping to form a tube for the flow of an imaginary fluid.

is . . . a purely imaginary substance," a device by which to see how the lines of force could determine the force (box 2.3).

Lines of force have sometimes been considered real. Here's the English physicist J. J. Thomson (1856–1940, winner of the 1906 Nobel physics prize for finding experimental evidence of the electron):

> [T]hese lines of force are not merely geometrical figments but . . . they – or rather the groups of them forming tubes of force that end on an electron – are physical realities.   (1925: 20)

More frequently, however, lines of force are interpreted as merely convenient devices for picturing the field. To see why, we must examine a phenomenon that was crucial in leading Faraday to emphasize

---

Box 2.3
Maxwell's device: an example

Imagine a positively charged body all by itself. It is the start of electric lines of force, so we are to imagine it as emitting fluid continuously at a rate proportional to the charge's magnitude. Consider the region shown in figure 2.10. Its "front wall" consists of part of the surface of a sphere of radius r, while its "back wall" consists of part of the surface of a

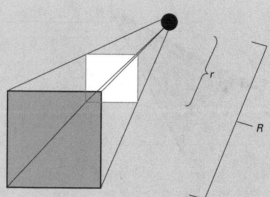

Figure 2.10 The imaginary fluid flows in all directions out of the positively charged body. Some flows into the sector through its front wall (at a distance *r* from the body). It then flows out of the sector through its back wall (at a distance *R* from the body).

sphere of radius *R*. Fluid from the source passes into the region through its front wall and out of the region through its back wall. For the fluid density in the enclosed region to remain unchanged, the quantity of fluid moving in during one second must equal the quantity moving out during one second. The fluid's density (ρ) multiplied by the front wall's surface area (proportional to r^2, since the surface area of a sphere of radius *r* is $4\pi r^2$), multiplied by the fluid's speed *v* crossing the front wall, multiplied by 1 second equals the quantity of fluid flowing into the region in 1 second. The quantity of fluid flowing each second out of the region through the back wall equals the corresponding product, proportional to ρVR^2 where *V* is the fluid's speed crossing the back wall; recall that ρ is the same at both walls since the fluid is incompressible. Since the same amount of fluid crosses both walls in one second, we have

$$\rho v r^2 = \rho V R^2, \qquad (2.8)$$

and hence

$$R^2 / r^2 = v / V. \qquad (2.9)$$

So the fluid's speed – and therefore the intensity of the electric force exerted by the charged body – is inversely proportional to the square of the distance from the charged body. Maxwell's device thus yields Coulomb's law (equation 2.1).

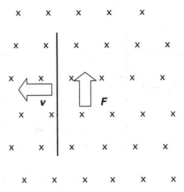

Figure 2.11 A uniform magnetic field, with lines of force pointing into the page (shown as ×s), occupies a region containing a metallic rod moving at velocity **v**. The force **F** on the electrons in the rod (in accordance with **F** = (q/c) **v** × **B**) is shown.

lines of force: the production of an electric current by magnetic forces ("electromagnetic induction").

In the next few pages, we'll look at quite a few experimental arrangements. Don't rush; make sure you understand each before going on to the next. That way, you will appreciate the moral at the end of the story. Let's begin with the simplest case. Suppose a rod that is metallic (and so electrically conductive) is moved with a steady speed and direction (velocity **v**) through a uniform magnetic field **B** caused by some external source, in the absence of any external electric field (figure 2.11). Each electron in the metal experiences a magnetic force directed along the rod. Electrons that are free to move will therefore flow along the rod. Static negative charge (from excess electrons) will accumulate at the end towards which current flows, leaving a positive charge at the other end. The flow will cease when the electric force on the electrons, repelling them from the negative end and attracting them to the positive end, balances the magnetic force attracting them to the negative end. Faraday hypothesized that the current is induced because the rod, in moving, "cuts" (that is, "sweeps across") magnetic lines of force (Faraday 1839: 74). If the rod is stopped, it ceases to cut any lines, and so a magnetic force is no longer induced. On this view, lines of force must be real since motion across them *causes* forces.

Now suppose the ends of the moving rod maintain contact with a pair of long wires that provides a path by which the electrons can return to their starting positions (figure 2.12). Electrons will then flow

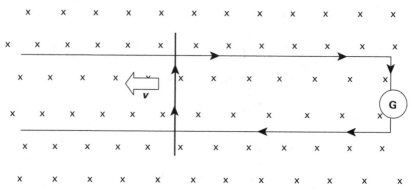

Figure 2.12 A uniform magnetic field, with magnetic field lines pointing into the page, occupies a region containing a metallic rod moving at velocity **v** in sliding contact with a return path for the current induced. A galvanometer G records the current; arrows indicate the flow of electrons around the circuit.

continually around the circuit. The long wires forming the return path are stationary. We can attach a galvanometer to the circuit to measure the current induced as the rod moves, cutting magnetic lines of force.[5]

Instead of the magnetic lines of force being stationary and the rod moving among them, we could have the rod stationary and the lines moving across it. Imagine two straight parallel wires (called "A" and "B"), each at rest and neither containing a current. Suppose a current is then switched on through A. Magnetic lines of force, encircling A, spread out from A like ripples on a pond (figure 2.13). As these lines of force sweep across B, a current is induced to flow through B. As Faraday puts it:

> [T]he magnetic curves themselves must be considered as moving (if I may use the expression) across the wire [B] under induction, from the moment at which they begin to be developed until the magnetic force of the current [through A] is at its utmost; expanding as it were from the wire [A] outwards, and consequently being in the same relation to the fixed wire [B] under induction as if it had moved in the opposite direction across them, or towards the wire [A] carrying the current.   (1839: 68)

Once the current in A stops rising and reaches a steady level, no additional lines of force emanate from A, and so shortly thereafter, no more lines sweep across B; the lines remain stationary in space, encircling A. Since B is no longer cutting any lines, B's current disappears.

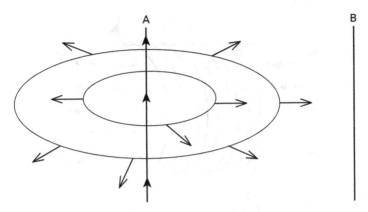

A   B

Figure 2.13   Electrons are made to flow down wire A in the direction of the arrows. Magnetic field lines looping around A emanate from the wire when the current in A is turned on. Arrows depict the spreading out of the magnetic field lines from A. Some of these lines are about to sweep across wire B, through which a current will then be induced.

If we decrease the current in A, magnetic lines of force move from the surrounding space back into A, crossing B and thereby again causing a current to flow in B.

Faraday pictured these lines as real causal actors. To see why lines of force are today usually considered unreal, we must examine how lines of force would account for some other examples of electromagnetic induction. Consider a "solenoid," a kind of electromagnet consisting of a wire coiled around a tube (figure 2.14). Suppose that the solenoid is carrying a steady current. If the solenoid is very long, then there is practically no magnetic field outside the tube (except near its ends). In the space enclosed by the tube, there is a uniform magnetic field pointing along the tube's length. In other words, the "density" of magnetic lines remains the same throughout the region enclosed by the tube, and the lines there run straight down the tube. Suppose that as before, a metal rod is moved at a constant speed through the region with uniform **B** while maintaining contact with a pair of long stationary wires that provide any induced current with a return path (figure 2.15). The rod cuts magnetic lines of force, inducing a current. The same applies if we do not slide the rod along the stationary wires, but instead move the entire circuit as a unit. Then the only part of the circuit that is cutting lines of force is the rod; the horizontal wires do not sweep across any lines. So if we keep the rod stationary but slide the rest of the circuit underneath it, no current is induced.

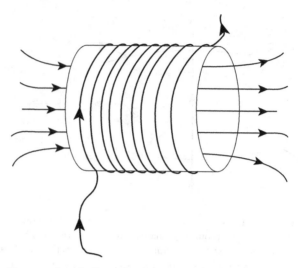

Figure 2.14   Here are the windings of a short solenoid; it could have many more. The arrows along the windings show the flow of the electrons forming the current. A few of the magnetic field lines are shown running through the tube and leaking out the ends. If the solenoid is long enough, **B** outside the solenoid is negligible near the middle of the solenoid's length.

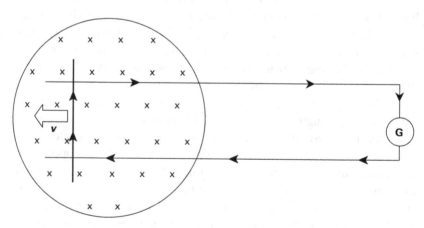

Figure 2.15   A uniform magnetic field created by a solenoid (projecting above and below the plane of the page), with magnetic field lines pointing into the page, occupies a region containing a metallic rod moving at velocity **v** in sliding contact with a return path for the current induced. A galvanometer G records the current; arrows indicate the flow of electrons around the circuit. The wires pass through gaps between the windings of the solenoid.

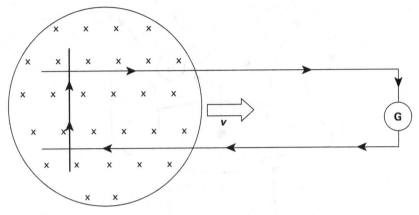

Figure 2.16   Similar to figure 2.15, except that now the solenoid is moving to the right at speed *v* rather than the rod moving to the left at speed *v*.

On the other hand, suppose that instead of moving some or all of the circuit while keeping the solenoid stationary, we keep the entire circuit stationary while moving the solenoid with a constant speed (figure 2.16). As long as the rod remains in the space enclosed by the tube, a current is induced. The explanation, in terms of lines of force, is that as the solenoid moves, the lines move with it, and so are cut by the rod (just as in the example involving wires A and B). The lines of force must be attached rigidly to the solenoid. This is confirmed by moving the solenoid and rod at the same speed and in the same direction, leaving the rest of the circuit stationary: no current is induced.

So far, so good for lines of force as real causal actors. But now consider a different arrangement. "Faraday's disk" (an experiment reported by Faraday in 1832) involves a metallic disk attached to one end of a cylindrical magnet (figure 2.17). The magnet's lines of force pass through the disk. The disk can be rotated while the magnet is held still, and vice versa. Wires running through a galvanometer touch the disk (at its center and edge) so that as the disk rotates, contact can be maintained (like a phonograph needle), different points on the disk's rim coming into contact with the outer galvanometer wire.

When the disk rotates while the rest of the apparatus is at rest, a current is induced, as measured by the galvanometer. This is easily explained in terms of lines of force: as the disk turns, the part of the disk between the galvanometer contacts (which is changing from

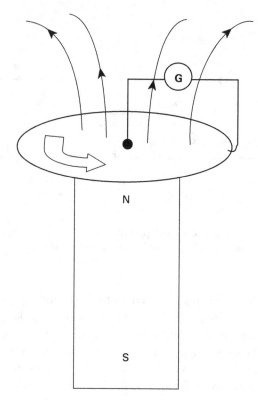

Figure 2.17   Faraday's disk. A few magnetic lines of force emanating from the magnet's north pole are shown. An arrow depicts the direction in which the disk is turning.

moment to moment) cuts lines of force, like the rod in the earlier example. But now something unexpected: if the magnet and disk are rotated together, so that the disk is *not* rotating relative to the magnet, there *is* a current. How can this be explained in terms of lines of force? If (as we saw earlier) the lines of force are attached rigidly to the magnet, like quills on a porcupine (Carter 1954: 168), then we would expect there to be *no* current: since the magnet is moving with the disk, the disk would not be cutting any lines of force. That the lines of force do *not* turn with the magnet is confirmed by rotating the magnet while keeping the rest of the apparatus still. No current is induced. If the lines were rigidly attached to the magnet, they would turn and so be cut by the part of the (stationary) disk between the galvanometer contacts, inducing a current to flow.

In sum, a current is induced if and only if the disk rotates – regardless of whether the magnet rotates. This fact persuaded Faraday that the lines of force of a rotating magnet "must not be considered as revolving with the magnet, any more than the rays of light which emanate from the sun are supposed to revolve with the sun" (1852: 31; also 1839: 63; 1935: I, 402).

How can the lines of force be rigidly attached to the magnet (like a porcupine's quills) for the purposes of straight-line motion but not for the purposes of rotation? This seems to involve a contradiction. If the lines of force are real things rigidly attached to the magnet, so that the magnet carries them along when it moves in a straight line, then the magnet must also carry them along when it rotates. Imagine a rotating porcupine. (Bet a book has never asked you to do *that* before!) So there does not seem to be a consistent way to regard lines of force as real.

But wait: In all of these arrangements, the wires attaching the galvanometer to the disk are stationary. If the magnet rotates and the lines of force turn with it, they sweep across the galvanometer wires. So even if the disk rotates with the magnet, the lines of force are still being cut by a conductor – namely, by one of the galvanometer wires – inducing a current. In this way, lines of force rigidly attached to the magnet can account for the induced current. What if the disk is stationary and the magnet turns? No current is induced. How can this be explained? The lines of force, rigidly attached to the magnet, are cut by the disk *and* by one of the galvanometer wires. No current flows because the force induced in the galvanometer wires urges the free electrons to flow around the circuit in one direction, while the equal force induced in the disk urges electrons to flow in the opposite direction. The two forces cancel each other. To see this more clearly, notice that for the same reason, no current flows when an entire circuit is moved in a straight line through a uniform magnetic field. In figure 2.18, segments AD and BC cut lines of force. But no current flows because free electrons in the wire "want" to flow from D to A and also from C to B. Neither clockwise nor counterclockwise flow around the circuit will do both.

So we can retain a consistent picture of the lines of force as real things, rigidly attached to the magnet, if we take the galvanometer wires into account. This alternative to Faraday's picture was proposed in 1885 by a now largely forgotten British telegraphic engineer, S. Tolver Preston (1844–1917):

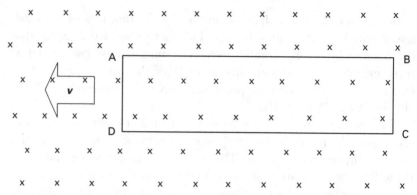

Figure 2.18   A rectangle of wire (with corners labeled A, B, C, and D) is moving at speed *v* to the left through a uniform *B* field with lines of force pointing into the page.

> [I]f (in regard to the inductive effect) the lines of force must be re-garded as partaking of the motion of a magnet when it is *translated* [that is, moving in a straight line], they must partake of its motion when it is *rotated*. This, therefore, constitutes [a] theoretic objection to the view adopted by Faraday.... [T]he same [induced current in the Faraday's disk apparatus] would be equally consistent with the *opposite* supposition (viz. that the lines of force partake of the rotatory motion of the magnet, in the same sense as they partake of the translatory motion of the magnet). For, admitting that the lines of force revolve with the magnet, then *they will intersect the galvanometer-loop circuit* when the magnet revolves on its axis; and this will evidently produce a current of the same direc-tion and magnitude as under Faraday's singular assumption, which he thought himself forced into by the experiment.   (1885: 133–4)

Obviously, Preston was concerned with developing a *consistent* picture of lines of force as real entities. (In chapter 7, we shall encounter another insight from Preston.)

But the inconsistency that Preston was trying to eliminate cannot be avoided so easily. Suppose we examine, not the current induced in a circuit, but the build-up of static charges at the ends of a rod. Then we do not need to detect a current and so can dispense with the galva-nometer (whose wires gave Preston a way out). Now we have a more direct way to examine whether the lines of force are attached to a rotating magnet just as they are to a magnet moving in a straight line. Take a cylindrical magnet surrounded by two metallic cylinders on the same axis as the magnet (figure 2.19). A metallic rod connects the two

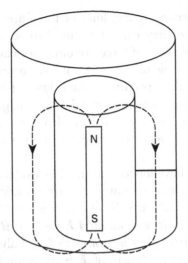

Figure 2.19 Two electrically conductive cylinders centered on a cylindrical magnet (with its north and south poles labeled). The cylinders are connected by a conductive rod. Dashed lines are examples of the magnet's lines of force.

cylinders. The magnet can be turned while the cylinders and rod are stationary, or vice versa. If free electrons are induced to flow down the rod, the cylinder into which they flow will become negatively charged, and the cylinder out of which they flow will become positively charged. The rod can then be broken (even while the rotation continues) and this charge difference measured. Essentially this experiment was performed on several occasions early in the twentieth century (Miller 1998: 146–50, 276–80).

Of course, if we keep the magnet stationary while the cylinders and rod rotate, then a charge builds up on the cylinders, since the rod cuts lines of force. But the key question concerns what happens when the magnet rotates: do the lines of force rotate with it, like the quills on a porcupine? It turns out that if we rotate the magnet while the cylinders and rod remain stationary, then *no* charge accumulates on the cylinders. Apparently, the lines of force (like the Sun's rays) do not rotate with the magnet.[6] This is confirmed if we rotate the magnet *together with* the cylinders and rod, so that they do not rotate relative to each other. Charges build up on the cylinders, so they must be turning through lines of force. Therefore, the lines of force would have to be rigidly affixed to the magnet for the purposes of straight-line motion but not for the purposes of rotation. This is an inconsistent picture.[7]

Because of this inconsistency, lines of force are generally believed to be unreal, and so they cannot be used to secure spatiotemporal locality. Nevertheless, lines of force are often presented in textbooks as a convenient way to picture a field, and electromagnetic induction is often characterized as occurring when lines of force are cut by a conductor. Of course, elementary textbooks can get away with this, since they avoid the cases where this idea leads to inconsistencies. Unfortunately, it then becomes unclear why lines of force cannot be considered real.

The examples we have looked at can all be explained in terms of the reality of fields. The two relevant laws are equation (2.7) governing magnetic forces ($\mathbf{F} = (q/c)\ \mathbf{v} \times \mathbf{B}$) and the modern form of Faraday's law of electromagnetic induction: $\mathit{curl}\ \mathbf{E} = -(1/c)\ \mathbf{B}'$. This says that a changing magnetic field (that is, non-zero $\mathbf{B}'$) is always accompanied by an electric field. (The "curl" of $\mathbf{E}$ is a vector function of $\mathbf{E}$ the details of which need not concern us now – but see discussion question 3 at the close of this chapter.) The changing magnetic field does not *produce* the electric field. (Recall that fields act only on particles, not on each other.) The two fields are joint effects of a common cause: the motions of certain charged bodies (Roche 1987; Rosser 1997: 124–6).

Let us see how these two laws apply in the examples that we initially tried to account for in terms of lines of force. For the rod moving in a uniform magnetic field (figure 2.11), $\mathbf{v}$ is at right angles to $\mathbf{B}$, so free electrons in the rod feel a magnetic force ($\mathbf{F} = (q/c)\ \mathbf{v} \times \mathbf{B}$) leading them to flow until this force is balanced by the electrostatic forces exerted by the charges accumulated at the ends of the rod. Next consider the two straight parallel wires at rest (figure 2.13). Wire B's $\mathbf{v} = 0$, so (by $\mathbf{F} = (q/c)\ \mathbf{v} \times \mathbf{B}$) it feels zero magnetic force. The increased current in wire A increases the magnetic field $\mathbf{B}$ in the region around wire B, so (by $\mathit{curl}\ \mathbf{E} = -(1/c)\ \mathbf{B}'$) an electric field $\mathbf{E}$ is created there, causing a current to flow in wire B. When we keep the circuit stationary while moving the solenoid with a constant speed (figure 2.16), the wire's $\mathbf{v} = 0$ while $\mathbf{B}'$ is non-zero at any location that the solenoid's field is just reaching or leaving behind. So at these locations – which coincide with the solenoid's own windings – there is an electric field, by Faraday's law. As a result, the top half of the solenoid becomes electrically positive, the bottom half negative, and the resultant electric field induces the rod's free electrons to move upward, initiating a clockwise flow of electrons in the circuit (Becker 1964: 372–3).

Now take Faraday's disk (figure 2.17). When the disk is rotating, then whether or not the magnet also rotates, the disk's electrons feel a force $F = (q/c)\, v \times B$; their v is at right angles to the magnet's B. On the other hand, if we rotate the magnet while keeping the rest of the apparatus still, then $B' = 0$ and the disk and galvanometer wires have $v = 0$, so no current is induced. Analogous remarks apply to the apparatus involving the cylindrical magnet surrounded by two metallic cylinders (figure 2.19). The magnetic field of the rotating magnet is nowhere changing (except because of minute accidental defects in the magnet). The lines of force will be unchanging on Faraday's view but moving on Preston's. Insofar as lines of force are mere representations of the field, they must be Faraday's lines, not Preston's.

I have just contrasted the field picture of various cases with the way they are understood in terms of lines of force. But in taking lines of force as real, Faraday was not favoring them over fields, since in Faraday's time, there was no concept of a "field" other than Faraday's notion of lines of force. In regarding these lines as real, Faraday was favoring them over action at a distance, not over fields. Although Faraday's lines of force are unreal, the field's reality would vindicate Faraday's instincts.

Discussion Questions

You might think about . . .

1   (a) Consider the arrangement in figure 2.15. Suppose that the solenoid and the wires are moving at the same uniform speed to the left. According to Faraday's theory, would any current be induced? (b) Now consider the situation in terms of fields. For the free electrons in the wires, is v non-zero? So do they feel a magnetic force? Is B' non-zero anywhere? So is there an accompanying electric force on the wire's free electrons? How must those magnetic and electric forces compare?

2   If the electric potential is unreal, why is it so dangerous to touch a high-tension (that is, high-potential) power line? How could you be hurt by a mathematical fiction?

3 I have expressed the law governing electromagnetic induction ("Faraday's law") in "differential form": **curl** $E = -(1/c)$ B'. It is logically equivalent to the following principle (figure 2.20a). Take a path that forms a loop. Cut the path into short segments. For a given segment, take E at that segment's midpoint. Consider that E's component in the direction taken by the path at the segment's midpoint. Multiply that component's length by the segment's length. Do this for all of the segments of the path. Add all of the results. Now allow the segments to become shorter and shorter. The sum converges to a certain quantity: this is the magnitude of E's "circulation" along the path. Now consider a surface bounded by that path (figure 2.20b). Cut the surface into small regions. For a given region, take B at that region's center. Consider that B's component perpendicular to the region. Multiply that component by the region's area. Take the result to be the length of a vector perpendicular to the region. Do this for all of the regions forming the surface. Add all of the vectors, and then allow the regions to become smaller and smaller. The sum converges to a certain vector: B's "flux" over the surface. The differential form of Faraday's law is logically equivalent to the principle that the magnitude of E's circulation along a path equals $(-1/c)$ times the rate of change of B's flux over the surface enclosed by that path. This is the "integral form" of Faraday's law.

Now consider a long solenoid. To a sufficiently good approximation, $B = 0$ outside the solenoid and B is uniform within, directed down the solenoid's length. Suppose that for a long time, we have been steadily increasing the current through the solenoid's windings. Accordingly, B is steadily increasing inside the solenoid. Consider a loop of wire around the solenoid's exterior (figure 2.20c). Since B is increasing at a constant rate inside the solenoid, B's flux over the surface enclosed by the wire loop is increasing at a constant rate, and so by the induction law, E must have an unchanging non-zero circulation around the loop. Hence, a current is induced to flow around the loop.

What causes the current? The cause is E's values at the loop (presuming fields to be real). But what causes E to take on non-zero values there? Suppose we answer: the changing B – the induction law specifies how a changing B causes a non-zero E. But B is *not* changing *at the loop*. The loop is outside the solenoid, and so is located where $B = 0$ throughout. Only *within* the solenoid is B changing. So even admitting the reality of fields, which were supposed to support locality, we seem

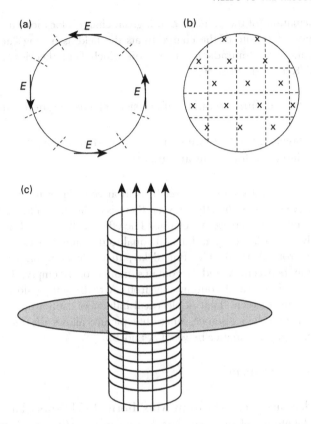

Figure 2.20
(a) ***E***'s circulation takes an especially simple form when ***E***'s magnitude is everywhere the same, the path is circular, and ***E*** is always directed along the path. (One scheme for dividing the path into short segments is shown by dotted lines.) The magnitude of ***E***'s circulation is the length of the path (the circle's circumference) multiplied by E.
(b) ***B***'s flux takes an especially simple form when ***B*** is uniform and always perpendicular to the plane of the surface (in this case, the page). (One scheme for dividing the surface into small regions is shown by dotted lines.) ***B***'s flux is the area of the region enclosed by the path (the circle's area) multiplied by ***B***.
(c) A solenoid (with a few of the windings and a few of the magnetic lines of force shown) surrounded by a loop of wire. The shaded area is the surface enclosed by the wire; we are interested in ***B***'s flux through that surface.

to have a violation of spatial locality: ***E***'s value at one place is caused by ***B***'s behavior at a distant place. This surprising result has sometimes been remarked upon in the physics literature:

The calculation [of the induced E at a given charge's location] using B values not at the site of the charge treats the field as having "action-at-a-distance" . . . contradicting tenets on which field theories . . . are based.   (Konopinski 1978: 502)

Is this the correct interpretation of classical electromagnetic theory?

4   Light travels through matter more slowly than through a vacuum. Consider, then, the following argument:

Imagine two point charges (A and B) held at rest, separated by some distance. Somewhere directly between them is a chunk of matter. Suppose we suddenly increase A's charge. The effect at B is not felt instantaneously; it is delayed by an interval equal to the time it takes for light to travel from A to B. The time-delay, then, depends on whether the region between A and B contains matter or is empty. Hence, some event must travel continuously from A to B, getting slowed by the matter in between. That is, electrical influences cannot operate by retarded action at a distance, since that picture allows the influence to "jump over" any matter between A and B.

Is this a good argument?

5   Consider these remarks from Max Born (1954 Nobel laureate, a principal developer of quantum mechanics in the 1920s):

[E]lectromagnetic forces are never observable except in connexion with bodies. Empty space free of all matter is no object of observation at all. All that we can ascertain is that an action starts out from one material body and arrives at another material body some time later. What occurs in the interval is purely hypothetical, or, more precisely expressed, arbitrary. This signifies that theorists may use their own judgement in equipping a vacuum with . . . fields, or similar things, with the one restriction that these quantities serve to bring changes observed with respect to material things into clear and concise relationship.
   This view . . . is an approach to the ideal of allowing only that to be valid as constructive elements of the physical world which is directly given by experience, all superfluous pictures and analogies which originate from a state of more primitive and more unrefined experience being eliminated. From now onwards . . . we have the abstract "electromagnetic field" as a mere mathematical device for conveniently describing processes in matter and their regular relationship.   (1924: 189–90)

What interpretation of fields does Born advocate? What is his argument for it? What could he mean by "directly given by experience"? "Superfluous" to what?

Notes

1 I use "Gaussian CGS" units. In other units, the right side of Coulomb's law includes a constant involving ϵ_0 (the "dielectric constant [or permittivity] of free space") or μ_0 (the "permeability of free space").

2 However, some nineteenth-century physicists suggested that Mercury's anomalous behavior could be eliminated by taking gravity's retardation into account (Roseveare 1982).

3 I shall not address arguments aiming to show that insofar as *any* scientific theory mentions entities that are not directly observable, that theory should automatically be understood merely as a device for predicting and systematizing our actual and possible observations, or instead should simply not be believed true. Instead of discussing arguments that attempt to deal with all hypothetical unobservable entities, I shall judge the reality of various hypothetical entities (such as fields and energy) *individually*, case by case. Readers interested in issues raised by *wholesale* "scientific anti-realism" are invited to consult van Fraassen (1980).

4 Bridgman's view here does not arise from wholesale anti-realism regarding unobservables. Rather, Bridgman says that "we have a right to ascribe physical reality to" a posited unobservable exactly when "it is uniquely connected with other physical phenomena, *independent of those which entered its definition*. This last requirement, in fact, from the operational point of view, amounts to nothing more than a definition of what we mean by the reality of things not given directly by experience" (1958: 55–6). This criterion is not satisfied by the field.

5 We are not getting something for nothing here: energy is being added to the system by whatever outside power is pushing on the rod, enabling it to maintain a constant speed. The magnetic lines of force, in being cut, are turning the energy of the rod's motion (which ultimately comes from the outside power) into the energy of the electric current. Of course, in Faraday's time, it was not yet known that an electric current consists of moving electrons.

6 At least, this is the most obvious interpretation. But see Djuric (1975), Viviani and Viviani (1977), and Djuric (1979).

7 One detail I have omitted is the need for an earthed (i.e., electrically grounded) metal case between the magnet and the cylinders if the magnet is electrically conductive. As the magnet rotates, charges redistribute in

the magnet, and the magnet becomes the source of an electric field. If there were no shield between the magnet and the cylinders, the rod would be affected by this field: a current would flow down the rod, building up a charge on the cylinders, even in the key case when the magnet rotates and the cylinders do not. The charge redistribution inside a rotating magnet arises from the fact that the magnetic field inside the magnet is aligned with the magnet's long axis, and so the magnet's rotation moves its electrons at right angles to the field; by $F = (q/c)\ v \times B$, an electron feels a force toward the exterior surface of one of the magnet's long sides. Electrons move to these surfaces until the electrostatic repulsion from the surface balances the magnetic force toward the surface.

3
Dispositions and Causes

1 Introduction

In chapter 2, I mentioned three rival interpretations of remarks like "The electric field here now is 5 dynes per statcoulomb." According to the first interpretation, this remark should be understood *literally* – as aiming to describe the state here now of a certain real thing: the electric field. If there is a body here now, then the electric field here now (together with that body's electric charge) causes an electric force on the body. Spatiotemporal locality is upheld. According to one rival interpretation, the remark aims to describe something about the charged bodies that have existed in various distant locations at various earlier moments – namely, that if a certain arithmetic operation is carried out on their charges, their distances away, and so on, the result is 5 dynes per statcoulomb. On this non-literal interpretation, the expression "The electric field here now" does not refer to a real entity existing over and above the bodies that have existed. The force on an electrically charged body here now is caused by its charge and some events that took place earlier at distant locations. No complete set of causes occurs arbitrarily near to the force. Spatiotemporal locality is violated.

This chapter will focus on the other non-literal interpretation I mentioned in chapter 2, according to which "The electric field here now is 5 dynes per statcoulomb" is made true by the fact that a charged point body here now, if there were one, *would* feel an electric force of 5 dynes for each statcoulomb of its charge (see box 3.1). In other words, this location possesses an "iffy" property: *if* a 1-statcoulomb point body

---

Box 3.1
Fields interpreted in terms of how a test charge would behave

Some interpreters of classical electromagnetic theory (e.g. Rosser 1997: 6) have offered the following objection to defining "E here now is 5 dynes/statcoulomb" as "if there were a charged point body here now, it would feel an electric force of 5 dynes for each statcoulomb of its charge." If a charged body had been here now, it would have changed the charge distributions on nearby conductors, which would have changed E's value from what it actually was. Hence, these interpreters suggest, E here now should instead be defined as the *limit*, as the body's charge approaches zero, of the force per charge that a point body here now would feel. It might be objected that because of retardation, the charge distributions on nearby conductors would not react *immediately* to the test body's being here now. Still, the conductors could be arbitrarily near, and in any case, so long as the charged body arrives here by moving in from somewhere else, there will be time for nearby conductors to react to its approach.

Of course, we might imagine the force that a charged body would have felt had it been here now, having *jumped* here directly rather than having traversed a continuous path from somewhere else. But this supposition imagines a violation of the laws of nature. While such a thing can be imagined without inconsistency (as I shall point out in this chapter), it might seem odd to use an imaginary violation of the natural laws to define a concept (E) whose main significance derives from its role in natural laws. (Maybe it is even worse than odd: had there been no law requiring bodies to move in continuous paths, then perhaps other laws of nature would have been different – such as those governing electromagnetism!)

Another option is to think of this interpretation as defining E in terms of the force that would have been felt by a charged point body had it been here now *and* the surrounding charge distribution been no different. This supposition is logically consistent with the laws and does not invoke a limit.

---

were here now, it *would* respond by feeling an electric force of 5 dynes. Such an "iffy" property is called a "disposition."

There are many dispositions. A porcelain vase's fragility is bound up with the fact that if it were treated roughly, it would likely then break.[1] Among a sugar cube's dispositions is its water-solubility (at a given pressure and temperature): were the cube immersed in water (under those conditions), it would dissolve. A sheet of paper is flammable and foldable, a piece of chalk is brittle and holdable, and a copper wire is malleable and electrically conductive. In contrast, a body's size, shape, and molecular structure are *not* among its dispositions. They are "categorical" properties.[2]

We might mention an object's fragility to encourage others to handle it with care. This emphasizes an important point: the porcelain vase is fragile even while it is *not* being treated roughly and even if it is *never* treated roughly. An object's dispositions are the threats and promises it makes, which may go unfulfilled. Its dispositions are its *powers* to bring about certain things and its *potentials* to react in various ways to various stimuli (see box 3.2).

To say that a body is fragile is to say what *would* (likely) happen if it *were* treated roughly. It is to assert the truth of a certain special kind of if–then sentence – a *subjunctive conditional*. We use subjunctive conditionals all the time, even when we are not discussing dispositions. For example, we say things like "I don't know whether we have any more butter in the house. If there were any more, some would be in the freezer." Sometimes we describe how history would have gone had things been different in some particular respect. For example, had Lincoln decided at the last minute not to attend Ford's Theater on the night of April 14, 1865, then he would not have been assassinated on that date. This subjunctive conditional is called a "counterfactual" because it purports to describe what would have happened had a certain situation come to pass that is "contrary to fact" – in other words, that did not actually happen. The subjunctive conditional "If

---

Box 3.2
Two distinctions

The categorical/dispositional distinction is not the same as the intrinsic/relational distinction introduced in chapter 1. The property of having a (living) brother is non-intrinsic but categorical; there is nothing "iffy" about it. Spatiotemporal relations are categorical. A body's water-solubility is dispositional but intrinsic: it imposes no logical constraints on the rest of the universe. There does not have to be water anywhere anywhen. It suffices that the body *would* dissolve, *were* it immersed in water (Langton 1998: 116–20, 150–1).

the match were struck, it would ignite" means that *either* the match is struck and ignites *or* the match is not struck but the following counterfactual holds: Had the match been struck, it would have ignited.

Let us return to dispositions. Recall the explosion of the space-shuttle *Challenger*. Suppose we want to know why this tragic event occurred. At the moment it was launched, *Challenger* must have been disposed to explode upon launch. In other words, roughly speaking, it was then the case that were *Challenger* launched, it would respond by exploding shortly thereafter. Now here is the key question for us: Was *Challenger's* explosion *caused* by its being launched and its being at that moment (unbeknownst to Mission Control) disposed to explode upon launch? This question is key because of the parallel question concerning fields. Suppose the remark "E here now is 5 dynes per statcoulomb" should be interpreted as ascribing a dispositional property to a given spatiotemporal location. (Recall that E is a scalar quantity: the magnitude of $\boldsymbol{E}$.) Suppose a point body charged to 1 statcoulomb occupies that location. Here's the parallel question: Is the 5-dyne electric force on the body *caused* by the body's carrying 1 statcoulomb of charge and E at its location (interpreted as a disposition) being 5 dynes per statcoulomb? If so, then although $\boldsymbol{E}$ is merely a dispositional property of some spot in space, the spot's having that property can still be a *local* cause of the electric force on the body there.

However, it seems too easy to say that *Challenger's* explosion was caused by its being launched while disposed to explode upon launch. The alleged cause seems somehow "too close" to the effect. The explosion's *true* cause, we feel, is whatever it was about *Challenger* that *made* it disposed to explode upon launch. These features of its structure, such as a gap in some rubber insulation around an engine, involve its categorical properties. Its possessing these properties constitutes the *categorical basis* (or "ground") of its disposition to explode when launched.

Early in the development of modern science, many natural philosophers argued that to account for an object's behavior, it is inadequate to say that the object received a certain stimulus and was disposed to respond with that kind of behavior upon receiving that kind of stimulus (Heilbron 1982: 17–22; Hutchison 1991). For example, although we can *predict* that a body will dissolve from knowing that it is water-soluble and being immersed in water, its dissolving cannot have been *caused* by its being immersed in water and being water-soluble, since there is nothing more to an object's being water-soluble than that it would dissolve were it immersed in water. The body's dissolving was *caused* by the water's having a certain molecular structure and physical state (temperature, pressure, and so forth), the body's having a certain molecular structure, and there obtaining certain spatiotemporal relations (namely, the body's being immersed in the water). The causally relevant features of these events all involve categorical properties. To say that its dissolving was caused by its being water-soluble and immersed in water is a cheat – the cheap maneuver of duplicating as a disposition the behavior for which we wanted to account. As David Hume (Scottish philosopher and historian, 1711–76) said ironically of natural philosophers who propose such "causes":

> They need only say, that any phaenomenon, which puzzles them, arises from a faculty or an occult quality [that is, a corresponding disposition], and there is an end of all dispute and enquiry upon the matter.   (1978: 224)

For any kind of behavior (such as dissolving when immersed in water), there exists the property of being disposed so to behave (such as the property of being water-soluble). This behavior is said to "manifest" that disposition. As we shall see, if a given property is manifested by only a single, narrow sort of behavior (as solubility is manifested only by dissolving), then that property is "too close" to that behavior to qualify as causally relevant to it.

From this idea, we can derive an important lesson regarding spatiotemporal locality: If "$E = 5$ dynes/statcoulomb" is properly interpreted as describing a disposition, then E cannot be a cause of a charged body's feeling a certain force. Thus, only a *literal* interpretation of the remark concerning E portrays E as genuinely a local cause of the electric force on a charged body. In later chapters, we will examine arguments for and against a literal interpretation.

2 Dispositions, Categorical Bases, and Subjunctive Conditionals

Let us suppose we are correct in our suspicion that a disposition does not help to cause the behavior that manifests it. Nevertheless, if a body behaves in a certain way and we want to know why, it can be *informative* to be told that the body is disposed to behave in that fashion under certain specified sorts of circumstances. For example, suppose you slide a floppy disk into a computer and the computer fails to read the disk. You ask why, and you are told that the computer is disposed to do that to any disk.[3] This tells you something that you did not know before: that the computer is broken, so there may be nothing wrong with your disk. Nevertheless, you do not know what *caused* the computer to respond to your (or any other) disk in this way. That account must involve the computer's categorical properties. Likewise, though we already know that the body is feeling 5 dynes of force for each statcoulomb of its charge, we still learn something from being told that E at its location is 5 dynes per statcoulomb, even if E is merely a disposition. For instance, we learn that the body would have felt the same quantity of force even if its mass or other properties had been different, so long as its charge was the same. We are also informed that any other body at that location, so long as it carried the same charge, would have felt the same quantity of force.

Let us try to understand why we learn something from being informed of the disposition. As we saw earlier, the subjunctive conditional "Were the body treated roughly, it would then break" is automatically true if the body is, in fact, treated roughly and breaks. Hence, if we already knew that the body was treated roughly and then broke, we would learn nothing from being told of this conditional's truth. But the computer-disk example suggested that we *would* learn something from being told of the body's fragility. So the conditional "Were the body treated roughly, it would then break" cannot really be the definition of "The body is fragile."

From being told that the body was fragile, we learn that the body would *still* have broken had it been treated roughly *while* (for example) *the weather was different.* In other words, we learn that the body would still have broken had it been treated roughly in *any* circumstance in a certain range – that is, under any set of conditions that is "standard." In contrast, the fact that the body broke while being treated roughly under the actual conditions does *not* entail that it would still have broken under rough treatment in *other* conditions. That is something we learn from being told of the body's fragility.

Hence, the subjunctive conditional that defines fragility must refer to any set of "standard conditions" or to the absence of any "disturbing factors." Let us see another example where this comes into play. Suppose a diamond is locked inside a box to which explosive charges and a motion sensor are attached. The apparatus is designed so that were the box shaken roughly, the explosive charges would go off, breaking the diamond. Suppose that since the diamond is locked inside the box, the diamond would be treated roughly only from the entire apparatus being shaken roughly. So the conditional "Were the diamond treated roughly, it would then break" is true. If this conditional were the definition of "The diamond is fragile," then the diamond would have to qualify as fragile, which is obviously incorrect.

Of course, this example is highly artificial; the diamond is in a very peculiar situation. However, that is precisely the key: we all recognize that the diamond is not in "standard conditions" for assessing its fragility. We all recognize this because long ago, in coming to grasp what it means for something to be "fragile," each of us learned which conditions qualify as "standard" as far as fragility is concerned, just as part of learning the law of falling bodies ("All bodies falling to earth accelerate at 32 feet per second per second, in the absence of disturbing factors") is learning which conditions involve no "disturbing factors." Bear in mind that the supposition "Were the body to be treated roughly under any set of standard conditions" does *not* mean "Were the body to be treated roughly under any set of conditions in which a fragile object would break were it treated roughly." We could grasp what count as "standard conditions" without having already understood what it means for an object to be fragile, just as we could know what sorts of conditions count as involving "no disturbing factors" in connection with the law of falling bodies even if we do not know the rest of the law: the rate at which bodies accelerate while falling in those conditions. The law does not mean "Every body falling to earth – in conditions where a body falling to earth accelerates at 32 feet per second per second – accelerates at 32 feet per second per second." Unlike this trivial truth, the law of falling bodies needed to be discovered through observations. Part of discovering it was discovering how "standard conditions" should be defined for the claim about falling bodies to be accurate enough for the intended purposes. We had to learn empirically that being near to the North or South Pole, for example, should qualify as a "disturbing factor."[4]

Of course, what specific conditions qualify as "standard" in connection with fragility will depend on the laws of nature. For instance, we understand that in an imaginary, science-fictional universe where dynamite is *not* explosive, the diamond might be in standard conditions for assessing its fragility even while locked inside a box fitted with dynamite. Nevertheless, we can characterize the "standard conditions" in any possible universe as those in which there are no other factors operating, such as high explosives attached to the body, set to go off if the body is treated roughly.[5]

Let us return to our main task of trying to understand whether we were correct in suspecting that a disposition is barred from being a *cause* of the behavior manifesting it. If this suspicion is correct, then E at a given spot, interpreted as a disposition, cannot be a local cause of the electric force on a charged body at that spot. We suspected that the behavior manifesting the disposition was caused not by the disposition itself, but rather by the non-dispositional structure underlying and in some sense responsible for the disposition: its "categorical basis." (The categorical basis of E at a given spot, for example, might be the charges and locations of various bodies at retarded times; the categorical basis of *Challenger*'s being disposed to explode upon launch involved a gap in a rubber seal around an engine.) To go any further, however, we must think about a very controversial matter: the *relation* between a disposition and its categorical basis. Let me briefly mention several alternative conceptions of this relation.

Some philosophers (e.g., Place 1996) have suggested that a disposition (say, a given sugar cube's water-solubility) is *caused* by its categorical basis (such as the cube's molecular structure). However, this view leads to some curious results. Consider, for instance, a key that possesses the power to open a certain lock. That power is one of the key's dispositions; it involves the subjunctive conditional "Were the key turned after being inserted into the lock under any standard conditions, the lock would open." What is the categorical basis of the key's power to open the lock? Perhaps it consists of various intrinsic properties not only of the *key*, but also of the *lock*. Then the disposition's categorical basis is not found entirely at the disposition's own "location" (where the key is). Therefore, if the categorical basis *causes* the disposition, we would have action at a distance on the cheap. Let me dramatize this: In altering the lock, we would instantly be causing a change in the key (namely, in the key's power to open the lock) no matter how far apart the lock and key are. Surely, something this mundane should not count as genuine action at a distance! (This is like some examples of non-causal connections that we discussed in chapter 1.) So the disposition's categorical basis must not cause the body to possess the disposition.

The obvious response to this argument is to suggest that the categorical basis of the key's power to open the lock consists solely of various categorical properties of the *key*, since the key is the body possessing the disposition. But this reasonable suggestion does not avoid action at a distance on the cheap. If the key's disposition is caused by its categorical basis, then the key's categorical properties are insufficient to form a *complete* cause of the disposition. Some of the lock's categorical properties must also be causally relevant to the key's disposition. This re-opens the door to action at a distance on the cheap. In short, some of the lock's categorical properties are no less causally relevant to the key's disposition than some of the key's categorical properties are, even if the lock's categorical properties are not part of the disposition's categorical basis. So action at a distance on the cheap can be avoided only if neither the key's nor the lock's categorical properties help to *cause* the key's disposition.

Another suggestion regarding a disposition's relation to its categorical basis is that the disposition just *is* its categorical basis – merely picked out in a special way (Mumford 1998; Ehring 1999). To understand this suggestion, let us first notice that the same person can be identified in many ways: as the person named "Marc Lange," as the author of this book, or as the occupant of Room M340 in Savery Hall. Likewise, any property can be identified in many ways. For example, the property of being 37 years old is the same as the property of being the same age as Marc Lange actually was on November 22, 2000. Now a sugar cube's water-solubility, on the view I am describing, refers to those of the cube's categorical properties that play a certain causal role: those that would cause it to dissolve in water. We may not know which categorical properties these are. But we can still refer to them indirectly, via one of their causal roles. If a body's disposition is identical to that disposition's categorical basis, then since the basis helps to cause the behavior manifesting the disposition, it follows that the disposition helps to cause that behavior.

What should we say regarding this view? Undoubtedly, we can pick out certain of a body's categorical properties by referring to them as the properties that give the body a certain disposition. However, the idea that the body's possessing some disposition is *identical* to the body's possessing certain categorical properties leads to some curious results. It seems obvious that the same disposition can have different categorical bases in different things. For example, objects can be unliftable (by human beings, unassisted by any special equipment) for many different reasons: one object may be too heavy to lift, another may be lighter but too bulky or too awkwardly shaped, another may be too slippery to grasp, and so on. Likewise, sugar and salt are both water-soluble, but for different reasons, arising from the fact that the atoms in a sugar molecule are linked by covalent bonds, whereas salt is an ionic compound. Thus, if a disposition is just the categorical basis picked out by its causal role, then a sugar cube's water-solubility must be a *different property* from a salt crystal's water-solubility. What the sugar and salt have in common is not water-solubility, but the property of possessing a property that would cause dissolving. This seems a bit strained.

In any case, this view of a disposition's relation to its categorical basis entails that if E is a disposition of a certain point in space, then although E there helps to *cause* the electric force on a charged body there, E is not a *local* cause. That is because on this view, "E here now" refers to various events involving distant charges at retarded times. So E at a given spot, interpreted as a disposition, cannot serve as a *local* cause of the electric force on a body there.

This takes me to what seems like the most plausible view of a disposition's relation to its categorical basis. On this view, a sugar cube's molecular structure (the basis), the relevant categorical properties of water (under standard conditions), and various laws of nature combine to logically entail the cube's water-solubility – to logically entail that were the cube immersed in water (under any set of standard conditions), it would dissolve. Laws establish the connection between the body's possessing the disposition's categorical basis (the sugar crystal's molecular structure) and the body's receiving the relevant stimulus (being immersed in water), on the one hand, and the body's manifesting the disposition (dissolving), on the other hand.

How does this apply to E? Suppose that E at a given spatiotemporal location, where there may be no matter, is 5 dynes per statcoulomb. If the basis of that 5 dynes per statcoulomb were some categorical property's being instantiated *at that same spatiotemporal location*, then since that location may contain no matter to possess categorical properties, something else would have to be there to serve as the categorical basis. That "something else" would be none other than a real field, existing on an ontological par with matter. We have suddenly stopped interpreting E as a disposition! To stick with that interpretation (which is our task in this chapter), the categorical basis of the 5 dynes per statcoulomb at the given spatiotemporal location would have to be *remote* in space and time from that location: various charges at retarded positions and times. Facts about those charges, combined with Maxwell's equations, logically entail that were a charged body at the given spatiotemporal location, it would feel 5 dynes of electric force per statcoulomb of its charge. If those distant charges cause the electric force on a body and E does not, then retarded action at a distance would have to operate.

So the only way for spatiotemporal locality to hold (while E is a disposition) is for E to help cause its own manifestation. But this is implausible: as we saw earlier in connection with the *Challenger* explosion, it seems that the disposition's categorical basis, rather than the disposition, does the causing. Of course (as we saw in chapter 1), there can be many complete causes of the same event, and the fact that one complete cause omits the disposition, in favor of its categorical basis, does not automatically preclude there being another complete cause that includes the disposition instead of its basis. However, there remains the feeling that the disposition is "too close" to its manifestation to count as a cause of it.

Let us try to refine this idea. The cube's water-solubility is *defined* as the truth of something like the following subjunctive conditional: were the cube immersed in water under any set of standard conditions, then it would dissolve. Therefore, if the cube's water-solubility combined with the cube's being immersed in water (under any set of standard conditions) to form a complete cause of the cube's dissolving, then this complete cause would *logically entail* its effect. In other words, it would be a *contradiction* for the cause not to be followed by this effect. But the link between a cause and its effect (or, more precisely, between instantiations of their causally relevant properties) is not as strong as a logical entailment; a cause brings about its effect only by the grace of the laws of nature. For example, the sugar cube's molecular structure (the disposition's basis), the relevant categorical properties of water (under standard conditions), and *the laws of nature* combine to logically entail that were the cube immersed in water (under any standard conditions), it would dissolve. The causes by themselves, without the laws of nature, are not enough to logically entail the effect.

The reason that a complete cause fails to logically entail its effect is because causes are linked to their effects by the laws of nature, but the laws of nature are not *logically necessary* truths. In other words, the same properties could *without contradiction* have figured in different laws of nature from those that actually govern the universe. For instance, instead of momentum being conserved in a collision between two billiard balls, the law governing collisions could have been that the faster speed is "contagious": that the ball that is moving slower before the collision (as judged from a certain special vantage point – say, the center of the universe) acquires the speed that the faster ball had before the collision, while the faster ball maintains that speed. (If the balls were moving equally fast before the collision, then the law could require that they both be stationary after the collision.) That the laws of nature are not *logical* truths also seems evident from the parlor game that physicists play of asking what the universe would have been like had the laws been different in some respect. The physicists' counterfactual suppositions ("counterlegals") would be contradictory if the laws were logically necessary. For example, had electromagnetic forces been ten times stronger, then the Coulomb repulsion between protons would have been 100 times stronger, and so all nuclei from carbon onward in the Periodic Table would have been unstable; the "strong" nuclear force would not have been strong enough to hold the protons together against their mutual electrostatic repulsion. So there would have been no stable carbon molecules, and hence no carbon-based life. This reasoning makes sense, even though it begins from a counterlegal supposition, because that supposition does not involve a contradiction, since the laws are true but not logical truths.

In short, the laws of nature help to determine what effects a given event has, and those laws are not logically necessary. Therefore, if C is a complete cause of E, there would have been no contradiction had C, with all of its actual causally relevant properties intact, brought about a different effect, inconsistent with E.[6] As Hume famously said regarding "the supposed tie or connexion between the cause and effect, which binds them together":

> When I see a Billiard-ball moving in a straight line toward another . . . may I not conceive, that a hundred different events might as well follow from that cause? May not both these balls remain at absolute rest? May not the first ball return in a straight line, or leap off from the second in any line or direction? All these suppositions are consistent and conceivable. . . . [T]he conjunction of [the actual effect] with the cause must appear equally arbitrary; since there are always many other effects, which, to reason [that is, to logic alone, setting aside the laws of nature – in other words, when we use only what follows logically from the causally relevant properties being possessed by the cause and effect], must seem fully as consistent and natural. (1977: 18–19)[7]

To predict the effects of a given cause, it is not enough to know the cause's categorical features and what they logically entail. We must also know the laws of nature, which we can ascertain only from observations of similar cases.

Hence, a disposition cannot help to cause the behavior manifesting it, since then a cause and its effect would be "too close": their connection would be logically necessary, not mediated by the natural laws. It is logically impossible – contradictory – for a water-soluble cube to be immersed in water (in standard conditions) and yet fail to dissolve. So the cube's dissolving cannot be *caused* by the cube's being immersed in water (in standard conditions) and the cube's being water-soluble (Mackie 1973: 136–8).

By the same reasoning, E (interpreted as a disposition) at a given spatiotemporal location cannot be a cause of the electric force on a body at that location. The only way to secure locality is for "$E = 5$ dynes/statcoulomb" to refer to a categorical property: the state of a field existing apart from and on a par with matter.

3 Are the Categorical Bases in Themselves Unknowable?

If "E here now equals 5 dynes per statcoulomb" describes a mere disposition of some spatiotemporal location, then the force on a charged body at that location must be caused by various categorical properties being instantiated elsewhere elsewhen, such as various quantities of electric charge being possessed by various distant bodies at retarded times. However, this raises an important question: Isn't a body's electric charge itself just one of its dispositions – the disposition, say, to exert a force on or to feel a force from another charged body? Electric charge might well seem like a disposition, considering that we know how a charged body would behave in various conditions and how its behavior would differ from that of an uncharged body in those conditions, but we do not know what electric charge is like *in itself*, what sets charged matter apart from uncharged matter. The same could apparently be said of any other basic intrinsic physical property.

But if a body's most basic properties are just dispositions, then (as we saw in the previous section) those properties cannot be causally relevant to their manifestations. Yet these most basic properties are supposed to be ultimately responsible for all manner of phenomena. For instance, a body's having a certain electric charge had better help to cause the electric force that the body feels. So these properties cannot be dispositions! They must be interpreted as categorical properties if science is to be interpreted as construing them as the fundamental causally relevant properties.

Nevertheless, we identify these properties not by what they are in themselves, but by what they do: by the dispositions they ground. Apparently, the term "1 statcoulomb of electric charge" merely names the whatever-it-is-we-know-not-what property that plays a certain causal role: were two bodies with this property long maintained at rest 1 cm apart, they would exert 1 dyne of electric force upon each other. The same applies to other properties that scientists construe as basic, such as a quark's "color." This is not color in the ordinary sense. A quark's "color" can be red, green, or blue, and an anti-quark's can be cyan, magenta, or yellow. The three quarks forming a proton, for example, must all be different colors according to the "Pauli exclusion principle," which dictates that no two quantum particles like quarks can be in the same state at the same time. So a quark's "color" is understood merely as some sort of respect in which it can differ from other quarks. Quarks of different colors also attract each other (like opposite electric charges), thereby holding together a proton. But we have no idea regarding what makes some quark "red" or how it differs in itself from a "green" quark.

If science eventually discovers how to understand electric charge (or a quark's "redness") in terms of more basic properties, then the original problem would simply apply to those more basic properties; they would just be new qualities that we do not understand in themselves. We have arrived, then, at the surprising conclusion that science cannot ever, even in principle, learn what the ultimate causes are like in themselves (Blackburn 1990; Broad 1933: vol. 1, 269–72; Foster 1982: 66–7; Langton 1998: 176; Mackie 1973: 150–2; Mumford 1998: 132–3; Price 1953: 322). Science can confirm that there is a certain categorical property ("possessing 1 statcoulomb of charge"), a body's possession of which produces certain effects and, in turn, is brought about under certain conditions. But if this property turns out to be basic, then as to what this property as such really is like: we can never know.

Likewise, suppose we give a literal interpretation to "The electric field here now is 5 dynes per statcoulomb." Then we ascribe to the electric field the property of having a strength here now of 5 dynes/statcoulomb. But what is involved in possessing this property? We have picked out this property in terms of its effect: for each statcoulomb of charge on a body here now, the body would feel 5 dynes of electric force. We have not identified the field's property through what it is in itself, independent of that causal role. As Hertz remarked:

> The interior of all bodies, including the free ether [what we would now call "space empty of ordinary matter"], can . . . experience some disturbances which we call electric and others which we call magnetic. We do not know the nature of these changes of state, but only the phenomena which their presence calls up.   (Born 1924: 161)

We might say that for the field to be 5 dynes/statcoulomb here now is for it to be *such that* a charged body here now would feel 5 dynes of electric force for each of its statcoulombs of charge. But this "such that" maneuver reveals how little we know about the field's property *as such* where it is capable of exerting 5 dynes of force on any statcoulomb of charge.

Therefore, it seems to follow that *either* electric charge and the other fundamental properties, according to our theories, should be interpreted as dispositions, in which case those theories fail to identify the fundamental causally relevant properties, *or* electric charge and so forth should be interpreted as categorical properties, in which case our theories fail to tell us about what the fundamental causally relevant properties are like in themselves. Neither of these options is especially attractive. So let's re-examine the above arguments to double-check that they really leave us with only these options.

Let us begin by trying to figure out whether or not electric charge and other such properties should be interpreted as dispositions. In the previous section, I explained that a property is dispositional in virtue of being defined in terms of a certain sort of subjunctive conditional. Notice the role of *definition* here. To be a disposition, it is not enough that the property be possessed by some thing if and only if the right sort of subjunctive conditional holds, nor is it even enough that this "if and only if" connection be "logically necessary" (that is, be such that its denial involves a contradiction). For example, a certain piece of paper's having four corners is intuitively one of its categorical properties. Yet this is a logically necessary truth: "x is four-cornered" if and only if "were x's corners counted correctly in any standard conditions, the result would be four."[8] Intuitively, this "if and only if" is not the *definition* of "x is four-cornered." The fact that an object is four-cornered is not *reducible* to some fact about what would result from its corners being counted correctly. On the contrary, that an object is four-cornered is *more basic* than – is responsible for – the fact about what would result from its corners being counted correctly. The point can be made even more blatant. This is a logically necessary truth: "*x* is four-cornered" if and only if "were *x* to travel at 80 miles per hour in any standard conditions, then *x*'s speed in miles per hour would be 20 times its number of corners." This is not the definition of "*x* is four-cornered."[9]

To determine whether electric charge is a disposition, we must get clearer on what it takes for the sentence "*x* possesses property *P*" to be *defined* in terms of some subjunctive conditional – what it takes, in other words, for the fact that *x* possesses *P* to be reducible to (to be analyzable in terms of, to be constituted by, to be nothing more than) the fact that a certain subjunctive conditional holds. Let us look at an example.[10] Suppose that temperature claims, such as "*x* has a temperature of 30°C," were introduced in science several centuries ago by being defined in terms of what would happen to a device of a certain construction (let's call it a "thermometer") were *x* in contact with it (in standard conditions). At this early stage in the development of physical science, "*x* has a temperature of 30°C" *meant* "were *x* immediately brought into contact with a thermometer (in any standard conditions), then the thermometer would respond by exhibiting behavior *B* [involving, say, the mercury reaching a certain level]." At that stage, any theoretical connection between "*x* has a temperature of 30°C" and other sentences was inherited from this subjunctive conditional; it was the *only link* between the temperature claim and the rest of science.

But not for long. Gradually, scientists discovered laws relating temperature to volume and pressure, to energy and specific heat, to solubility, to chemical reaction rates, and so on. Having discovered temperature's other diverse connections, scientists then encountered an anomaly: a body that behaved in every observed respect as if its temperature were 30°C (according to temperature's various connections) except that when the object was brought into contact with a certain thermometer (in what were understood to be standard conditions), the thermometer did not exhibit behavior *B*. Many possible explanations of this anomaly were considered. Perhaps scientists had made some mistake regarding temperature's other connections, or the body's temperature was not measured in standard conditions, or the thermometer was broken. The key point, however, is that the property of having a temperature of 30°C had by then acquired enough connections to other concepts that scientists could also intelligibly consider one further possibility: perhaps the body's temperature really was 30°C despite the body's failure to make a thermometer exhibit B in standard conditions. In particular, since heat flows from hot bodies to cold ones in contact with them until their temperatures are equalized, a thermometer placed in contact with (say) a much cooler body may not accurately register what that body's temperature is. The thermometer and body may reach equilibrium at a temperature non-negligibly higher than the body's original temperature.

This means of dealing with the anomaly involved abandoning the "if and only if" connection by which the notion of "temperature" had been introduced. However, by this point, the temperature concept had acquired so many and such diverse connections to other concepts that the loss or revision of any one of them would not have amounted to a *redefinition* of "temperature." Temperature's other connections were enough to hold the concept in place while the connection between temperature and the thermometer conditional was rejected. That connection was by this stage just one among temperature's many important connections, not the single special connection defining "temperature."

My point is that in rejecting this connection, scientists were not changing their *definition* of temperature, because by this point, there was no special class of connections holding purely in virtue of the meaning of "temperature." In other words, there was no *definition* of "temperature." This is not to say that "temperature" was meaningless (in the ordinary sense of the word) or that dictionaries could not come up with anything to say about it. Nor do I mean that a *host* of connections *collectively defined* "temperature." That would require me to say that whenever a single one of these connections was changed, the definition was changed, and so scientists were using a different temperature concept than before. Rather, once temperature acquired enough diverse important connections, the concept became too rich to have anything like a definition.[11] Though various connections differed in importance, there was no natural place to divide the connections into those that fixed temperature's meaning and those that did not.[12] None of the connections was logically necessary, since if any few of them had been different, the rest would have remained available to hold the concept fixed.

We should not interpret the scientists, in rejecting temperature's connection to the conditional initially defining it, as changing which property they were talking about as "temperature." Rather, they were discovering more about temperature, the same property that they had already been talking about. Of course, our example would be more realistic if the scientists did not simply *reject* temperature's original connection to thermometers, but rather amended the thermometer subjunctive conditional by including the thermometer's own initial temperature as a possible disturbing factor. But this amendment nevertheless would not have amounted to a change in the *meaning* of "temperature," as it would have had "*x* has a temperature of 30°C" still been *defined* in terms of the original subjunctive conditional concerning a thermometer.

In this example, "temperature" was initially defined in terms of a subjunctive conditional, and so referred to a disposition. Sometime before the anomaly was discovered, "temperature" had acquired enough diverse important connections that the original one was no longer *special* – that is, was no longer entirely responsible for giving "temperature" its meaning. So the meaning of "temperature" did change at some earlier point: when the thermometer conditional ceased to define "temperature." However, insofar as this example includes such a shift in meaning, it seems unrealistic. In reality, long before scientists discovered temperature's roles in various laws, scientists *expected* there to be some such roles and began trying to uncover them. So scientists did not regard temperature's connection to the thermometer conditional as privileged, even before any of temperature's other connections had been discovered. The property of having a temperature of 30°C was never thought to be *nothing more* than the disposition to affect a thermometer in a certain way. That conditional was *never* what "temperature" *meant*.

Contrast the property of having a temperature of 30°C with a genuine disposition. As I noted earlier, objects can be unliftable (by unassisted human beings) for many different reasons: their mass, their size, and so forth. Therefore, scientists do not try to generalize from their observations of *some* unliftable objects to laws governing *all* unliftable objects. An object's unliftability is nothing more than – is defined as – a certain subjunctive conditional's holding of it. Unliftability is a disposition.

As another example, consider the dispute among psychometricians over the concept of "general intelligence" (Sober 1982: 593–4). Some psychometricians believe that a person's doing well on IQ tests (in standard conditions) reflects her possession of a certain non-dispositional property (which they call "general intelligence"), just as a thermometer's reaction (in standard conditions) reflects a non-dispositional property (temperature). These scientists expect general intelligence to figure in many important (as yet undiscovered) psychological generalizations, linking it to various genetic and environmental factors, neurological properties, capacities to perform various tasks well, and so on. Other psychometricians believe that the capacity to do well on IQ tests (in standard conditions) is just a disposition. On this view, different people do well on IQ tests for different reasons, just as different objects are unliftable for different reasons. The capacities enabling a person to figure out a word problem quickly are unrelated to the capacities enabling a person to figure out a spatial puzzle quickly. A person's score on an IQ test is not sensitive to these differences, and so is not a reliable basis for predicting her other traits. A person's score on one IQ test suggests only that she would likely do well on other IQ tests. Whether the capacity to score well on IQ tests is merely a disposition, or there exists a single categorical property ("general intelligence") possessed by anyone who scores well (in standard conditions), remains an open question.

Because there is *nothing more* to a body's being fragile than a certain subjunctive conditional's holding, the vase's being fragile and treated roughly is linked by logical necessity to its breaking, and so the vase's fragility cannot be causally relevant to its breaking. A disposition's *specificity* to a single kind of behavioral manifestation is crucial to making a subjunctive conditional the disposition's *definition*. Thus, the reason a body's water-solubility cannot help to cause its dissolving is because of water-solubility's specificity to dissolving – exactly what many sixteenth- and seventeenth-century natural philosophers objected to in physical theories invoking "faculties," "powers," and the like as fundamental causes. As Isaac Newton (1642–1727 – aw, heck, I bet you've heard of him!) said in a famous, obscure, but nonetheless illuminating passage:

> These Principles [such as (I think he means) the categorical basis of a body's capacity to exert and to feel gravitational forces] I consider, not as occult Qualities, supposed to result from the specifick Forms of Things, but as [acting in accordance with?] general Laws of Nature, by which the Things themselves are form'd. . . . To tell us that every Species of Things is endow'd with an occult specifick Quality by which it acts and produces manifest [that is, observable] Effects, is to tell us nothing [as Hume said in the passage I quoted earlier]: But to derive two or three general Principles of Motion from Phaenomena, and afterwards to tell us how the Properties and Actions of all corporeal Things

follow from those manifest Principles, would be a very great step in Philosophy.   (1952: 401–2)[13]

Newton offered no theory to explain how masses are able to cause gravitational forces far away from them. Nevertheless, he emphasized that being 5 kg in mass is not merely the disposition to exert and to feel certain gravitational forces, since mass is linked to gravity *by natural law* rather than *by definition*. According to Newton, mass is not "specifick" to gravity, since besides the law of gravity, mass figures in other basic laws already known (such as Newton's second law of motion, linking a body's acceleration to its mass and the force on it), and mass was expected to figure in further basic laws as yet unknown, such as laws governing chemical reactions. (We will look at the concept of mass more closely in chapter 8.) With the same few fundamental categorical properties expected to be ultimately responsible for all natural phenomena, no single role played by any of these properties will be adequate to define it.

The same reasoning applies to electric charge. It is not merely the disposition to exert and to feel electric forces because it stands in other important relations. A body's charge influences the magnetic force it feels (equation 2.7). Charge is a conserved quantity, and laws dictate that a system's charge is the sum of the charges of its parts. Since electric current is the flow of charge down a wire, charge appears in various laws involving voltage, resistance, the heating of wires, and so on, as well as various chemical laws (such as Faraday's law of electrolysis). I am inclined to conclude that none of charge's important connections to various subjunctive conditionals is set apart from the rest as the one defining it.

Accordingly, we might say that bodies with the same electric charge have "something real" in common (unlike, say, bodies that are unliftable). On the basis of this conviction, we are willing to take a few observations of how some bodies behave when possessing a given quantity of charge, and to generalize from them regarding how any other, similarly charged body would behave, thereby arriving at various natural laws governing charge. But this raises the puzzle I mentioned earlier: if bodies with the same electric charge have *something* in common, then *what is it* they share? If charge (or mass, or any other fundamental categorical property) is just something-we-know-not-what, identified only as the categorical basis of certain dispositions, then we do not know what charge is *in itself*. Moreover, if we developed a better theory and thereby came to understand charge in terms of some more fundamental properties, the same problem would then afflict those properties.

But what would *count* as grasping electric charge (or any other property) *as such*? What knowledge are we supposedly missing?

Consider a property that we allegedly know in itself, such as the property (that a number might possess) of being prime. We know what it is for a number to be prime: to have no factors besides one and itself. But this property can be possessed only by numbers; it is not a *physical* property, one that might be causally relevant. So this property does not help us to understand what it would be to grasp electric charge in itself.

Let us, then, take a physical property: being square. It might be thought that we know what squareness *is*, not just the effects it has, because we have seen bodies being square. Unlike being square, a body's possessing a certain charge does not affect our senses directly. However, are we perhaps confusing knowing a property in itself with knowing what a body looks like (or sounds like, or tastes like) by virtue of possessing that property? It is very easy to confuse grasping a physical property in itself with being able to form a mental image of what it is to possess that property, where the raw materials for that mental image are derived from our senses (principally sight). Unless the property evokes a definite mental image, we tend to think that we do not understand the property as such. But this seems very narrow-minded. It seems more accurate to say that mental imagery can often mislead us, as when we picture a subatomic particle as a tiny ball, which we cannot imagine without coloring it in. Yet we all know that subatomic particles have no colors (Smart 1963: 74).

For that matter, why shouldn't we say that from the impressions that various bodies leave on our senses of sight and touch, we learn only that squareness is whatever property serves as the categorical basis for the disposition to affect our senses in these ways? Geometric properties, like size and shape, may initially seem to be ideal cases of properties we know in themselves. But insofar as these are *physical* properties, to be instantiated by matter in space and not merely by abstract mathematical entities, it is not obvious that our senses disclose to us these properties as such.

Without any obvious examples of knowing a physical property in itself, we must look elsewhere for a clear conception of what it would be to have such knowledge. Why do we think that there is something more to know about electric charge than what it *does* – namely, what it *is* for matter to be charged? The main reason, I suspect, is that we harbor the thought that if we grasped electric charge as such, then we would see why charged bodies are *logically compelled* to do what they do. In other words, we tend to think that the dispositions associated with being electrically charged (such as the power to exert a certain force under certain circumstances) are logically inevitable considering what it *is* for a body to be electrically charged, just as a prime number's being odd if it is greater than 2 follows logically from what it is for a number to be prime. In the words of John Locke (English philosopher, 1632–1704):

> The whole extent of our Knowledge, or Imagination, reaches not beyond our own Ideas, limited to our ways of perception. Though yet it be not to be doubted, that Spirits of a higher rank than those immersed in Flesh, may have as clear ideas of the radical Constitution of Substances [that is, of the fundamental categorical properties], as we have of a Triangle, and so perceive how all their Properties and Operations flow from thence: but the manner how they come by that Knowledge, exceeds our Conceptions.   (1975: 520)

These "Spirits," from their God's-eye viewpoint, know charge in itself, and thus "perceive" how the dispositions associated with being charged "flow" with logical inevitability from what it is to be charged.

However, this thought runs contrary to a view I defended in section 2: that the laws of nature are not *logically necessary*. The same categorical property could have served as the basis of different dispositions. On this view, there is a kind of arbitrariness in the laws of nature; certain laws are *basic* in that they have no explanation. (I shall discuss this idea further in chapter 4.) Nevertheless, it need not be arbitrary that, for example, the disposition to exert electric forces is always accompanied by the disposition to exert magnetic forces (under the proper circumstances), since this correlation might be accounted for by laws more basic than the electric-force laws and the magnetic-force laws. (Such turns out to be the case, as we shall see in chapter 7.) Often, science has reduced its estimate of the number of independent laws there are. Still, the truly basic laws are not compulsory on pain of contradiction. The cosmos is ruled by the laws of nature, not just the law of non-contradiction.[14]

I suspect that a grasp of the fundamental categorical properties "in themselves" would have to be nothing less than an explanation of why the basic laws must be as they are. Therefore, we should resist any temptation to think that there is such a mode of grasping the fundamental categorical properties. It is a mirage. There is no reason why the basic laws must be as they are.

It might be objected that for the laws to relate various properties to one another (as Coulomb's law relates being electrically charged, feeling a certain force, and being a certain distance apart), there must for each property be something definite that it involves in a way that is logically prior to all of the laws in which it figures (Armstrong 1968: 282; Kneale 1949: 94). Otherwise there would be nothing for the laws to relate: the laws would be establishing a mere formal network of relations, a flowchart connecting boxes none of which stands for anything! This objection leaves us facing a very difficult dilemma. If a categorical property is nothing more than a node in a network of relations, then there is nothing prior to that network for those relations to relate, which seems to make no sense. On the other hand, if a categorical property is something more than all of its relations to other such properties, then to grasp that "something more" would be to understand the categorical property in itself. But science could never supply that kind of understanding of the fundamental categorical properties.

My somewhat tentative suggestion is that we try to navigate between the horns of this dilemma. On the one hand, it is a mistake to think of the laws as relating properties having natures fixed prior to all of the laws. The laws help to *constitute* the properties they "relate." On the other hand, a categorical property *is* something more than its *particular* relations, in all of their detail, though that "something more" still involves nothing but the property's place in the network of relations. The laws could *all* be varied in some of their details without changing the properties to which they refer. For example, had Newton's gravitational-force law ($F = GmM/r^2$) instead portrayed the gravitational force F as varying with the inverse-cube of the bodies' separation r, and Newton's second law of motion ($\boldsymbol{F} = m\boldsymbol{a}$) instead related force $\boldsymbol{F}$ and mass m to the body's acceleration $\boldsymbol{a}$ squared, then mass would still have occupied the same causal role. That role is individuated by the laws it involves, but nevertheless by something over and above the details of those laws. Bodies would still have had masses – there would still have been a single parameter relating a body's motion to the force it feels – even if that relation had differed from $\boldsymbol{F} = m\boldsymbol{a}$.

Since we can characterize a categorical property independently of any one of the dispositions associated with it, a body's possessing that categorical property can help to *cause* behavior manifesting that disposition. A categorical property is more even than all of the dispositions associated with it, since contrary dispositions could have been associated with it. But we should resist turning this good thought into the bad thought that there is something more to possessing the categorical property, for ever unknown to us, knowledge of which would supply a deeper understanding of that property than knowledge of the laws would give us. There is nothing more; the fundamental categorical properties are not intelligible apart from their causal roles.

The same thoughts apply to the electric field. If E's being 5 dynes/statcoulomb here now is not merely a disposition of a spatiotemporal location, but ascribes a categorical property to a real thing, the electric field, then what is this property in itself? Indeed, what is the field in itself, a kind of ghostly immaterial whatnot pervading all of space? Don't we (as Hertz says) know only what it does, not what it is? No: its nature is given by the laws governing it. The field's properties figuring in those laws (such as its strength and direction at a given point) are not properties we sense directly or can picture in familiar terms. But that does not mean we fail to understand them fully. To ask "What does it mean for E to be 5 dynes/statcoulomb here now, other than that it would have various effects, such as causing a charged body here now to feel a certain force?" is to demand that we characterize what it is for E to be 5 dynes/statcoulomb *in isolation from* the relations established by natural law between this property and others. The question, in other words, has an answer only if the field's properties stand in interesting relations that are logically necessary. They do not, precisely because of their fundamental causal relevance. The field acts as it does not because it would be contradictory for it to do otherwise, but because that's just how the universe works.

Discussion Questions

You might think about . . .

1   Cooling a rubber band causes it to become brittle: were it stretched, it would break rather than return to its original length. What is the effect of cooling: the rubber band's acquiring the disposition (brittleness) or the rubber band's acquiring a certain categorical property (the disposition's basis), or both?

2   I have portrayed the contrast between dispositional and categorical properties as sharp. But perhaps some properties are both dispositional and categorical:

> When a bird is described as migrating, something more episodic is being said than when it is described as migrant, but something more dispositional is being said than when it is described as flying in the direction of Africa.   (Ryle 1949: 142)

Do you agree? Can you think of other examples?

3   Imagine the following excerpt from a newspaper:

> Researchers have confirmed what some parents have long suspected – that you cannot do a thing with some children's hair. And they believe unruly hair is caused by a little-known condition called uncombable hair syndrome.   (Hutchison 1991)

Consider, then, the following argument:

> Uncombable hair syndrome could be a genuine cause of unruly hair. After all, it assigns a responsibility for the unruly hair to the hair itself. This rules out parental self-delusion about the child's hair, atmospheric disturbance moving the hair, other children having ruffled it, a psychosomatic cause, the laziness of the child when it comes to combing, static electricity, and witchcraft. So since the appeal to uncombable hair syndrome may be false, it is not empty, and so may be a cause of unruly hair.   (Mumford 1998: 138–9)

Is this a good argument? What would it take for this syndrome to be a cause of unruly hair? When we cite gravity as causing an object to fall to the ground after being thrown skyward, is "gravity" just another name for "falling object syndrome"?

4   Suppose categorical basis B grounds disposition D. Then B has the power to bring about D. This power, in turn, needs a categorical basis B*, its being in virtue of B* that B grounds D. But then B*'s power to bring it about that B gives rise to D will itself need a categorical basis B**, and so forth infinitely. What is wrong with this argument (Blackburn 1990: 64)?

Notes

1   I'll not worry about whether various dispositions are "sure-fire" or merely "probabilistic."

2   Other terms sometimes used to express "non-dispositional" affirmatively are "occurrent," "structural," and "qualitative."

3   Alan Hajek used this example in a conversation with me some years ago.

4   Contrary to Martin (1994). See Lange (2000: 160–88).

5   I have cunningly used another disposition (being explosive) here. Bird (1998) and Martin (1994) give other sorts of cases that I believe can be reconciled with a counterfactual account of dispositions by appealing to standard conditions. Goodman (1983: 39) makes a similar appeal to standard conditions. What counts as rough treatment under standard conditions for assessing an object's fragility may be quite different for different kinds of objects: a weld on an airplane's wing, a bird's egg, the main supporting beam of an earthquake-damaged building. I shall set aside these and other complications.

6   I am not appealing here to the probabilistic character of some laws. The same applies even if all laws are deterministic.

7   For dissent, see Harre and Madden (1975: ch. 3).

8   Here "correctly" means "in a reliable manner," that is, "in a manner that would yield the truth about how many corners there are, no matter how many there were [perhaps within some understood range]." (Obviously, if "correctly" meant "so as to give the answer 'four'," then "were x's corners correctly counted, the result would then be four" would hold regardless of whether x is four-cornered.) Mellor (1974: 171) and Goodman (1983: 40–1) give similar examples, purporting to show that it is hard to distinguish dispositions from categorical properties.

9   In "standard conditions," x's going at 80 mph does not change its number of corners from their actual number.

10 I borrow the kernel of this example from Hempel (1978: 139–41). The lesson I draw directly from this example is also Hempel's (though my general picture of a disposition's limited explanatory role is not). Some philosophically minded readers might be surprised to find an "old-fashioned logical empiricist" like Hempel making a point so reminiscent of Quine.

11 Diversity is important. Some dispositions are commonly called "multi-track" to indicate that there is more than one subjunctive conditional to which they are connected by logical necessity. For example, a *hard* object would make a sharp sound if struck, would resist if pressed against, would result in a painful bruise on a person who banged against it, and so forth. But these are not diverse; they share a core idea. (That's why we understand what would count as "so forth.") So "hard" has a definition.

12   This account of temperature is in some respects similar to Putnam's notion of a "law-cluster" concept:

> Law-cluster concepts are constituted . . . by a cluster of laws which, as it were, determine the identity of the concept. The concept "energy" is an excellent example of a law-cluster concept. It enters into a great many laws. It plays a great many roles, and . . . in general, any one law can be abandoned without destroying the identity of the law-cluster concept involved.   (1975: 52)

13   To grapple with this passage further, see McGuire (1968).
14   Echoing Kant (Langton 1998: 174).

4

Locality and Scientific Explanation

1 Is Action at a Distance Impossible?

How can we discover whether spatiotemporal locality is satisfied? Some natural philosophers have suggested that one thing cannot *possibly* affect another without touching it (or touching something else that touches something else that . . . that touches it). For example, in thinking about how gravity works, Newton wrote:

> [T]hat one body may act upon another at a Distance thro' a Vacuum, without the Mediation of anything else, by and through which their Action and Force may be conveyed from one to another, is to me so great an Absurdity, that I believe no Man who has in philosophical Matters a competent Faculty of thinking can ever fall into it.   (Cohen 1978: 302–3)

Newton's thought here sounds very much like the idea behind the Russian/Jewish joke with which I began chapter 1: for a body's influence to be felt at a distant location, something must carry that influence through the region between the body and that location. Admittedly, this idea is very commonsensical. But we want to know *why* action at a distance is impossible. If we are told that it is impossible because a body's influence cannot arrive at a distant location without being carried through the region in between, then we still don't know why the body's influence must pass through the region in between in order for it to arrive at the distant location. The reply "Because action at a distance is just absurd" takes us right back to where we started; it amounts to nothing more than name-calling.

Perhaps Newton had in mind an argument like this, offered by Thomas Hobbes (English philosopher, 1588–1679, perhaps best known today for his views on government):

> [If a body] shall be moved, the cause of that motion . . . will be some external body; and therefore, if between it and that external body there be nothing but empty space, then whatsoever the disposition be of that external body or of the patient itself, yet if it be supposed to be now at rest, we may conceive it will continue so till it be touched by some other body. (Hesse 1965: 113)

This sounds reasonable enough: a body at rest will remain at rest unless touched by another body. This combines the principle of inertia with spatiotemporal locality. But can you identify Hobbes's *argument* for this idea? It seems to be that for one body to cause another to start moving, they must touch because if they are separated by a gap, then one cannot affect the other. But this argument is *circular*: it derives spatial locality from the assumption of, well, spatial locality. What we are interested in finding is an argument deriving locality from ideas that seem far enough away from locality to make the derivation worth giving – that is, far enough away for them to have their own sources of plausibility, independent of locality's.

This is difficult to do. Here is another argument concerning gravity, this time given by Samuel Clarke (English metaphysician and theologian, 1675–1729):

> That one body should attract another without any intermediate means, is indeed . . . a contradiction: for 'tis supposing something to act where it is not. (Alexander 1956: 53)

However, that a body acts where it is not (action at a distance) leads to a contradiction only when combined with the assumption that a body is present wherever it has an effect (spatial locality). Together these entail that a body is present at a location where it is not present. That is a contradiction! But to get it, we had to begin by presupposing locality. That is no way to argue for locality.

2   Brute Facts and Ultimate Explanations

Perhaps some of our sympathy for locality is prompted by this idea: If a cause were distant in space or time from its effect, there would be no explanation of its power to bring about its effect. This idea presupposes that to account for an effect, we need to explain it in terms of one thing touching another. To make this presupposition is really just to assume spatiotemporal locality from the outset.

However, a more refined version of this argument is not so obviously circular. Various laws govern the operation of gravity, electricity, magnetism, and other forces. For instance, gravitational and electrostatic forces diminish with the square of the distance from the body exerting them. If these forces operate by some local process, then the details of that process presumably explain why these forces diminish with the square of the distance. However, if these forces involve action at a distance, then no process is available to account for the specific character of the laws governing them. Apparently, those laws could just as well have been different; there is no reason why the forces could not instead have diminished with the cube of the distance, for instance.

Of course, to say that some local process is responsible for gravitational and electromagnetic forces is a far cry from describing that process, and without that description, we cannot explain the details of those force laws. Nevertheless, it might be insisted that *there must be* explanations of these force laws, even if we don't know what they are. For instance, as we saw in chapter 2, the electric effect of a distant charged body is delayed by an interval of time equal to the body's distance divided by about 300,000,000 meters per second (the speed of light). Why is it 300,000,000 m/s rather than some other speed? If there is some medium carrying the influence from the cause to the effect, then the speed presumably depends on certain physical characteristics of that medium, and these same characteristics presumably also explain other details of Maxwell's equations. Likewise, the fact that this medium occupies a three-dimensional space has sometimes been thought to explain why the force diminishes with the *square* of the distance (since a sphere's surface area increases with the square of its radius, and perhaps the stuff carrying the electric force out in all directions from a charged body spreads out evenly over all of the points at a given distance from that body, which form a spherical surface centered on that body). On the other hand, if these forces resulted from retarded action at a distance, then apparently, the extent of the time delay and the inverse-square character of the force would have to be "brute facts": although these facts could have been different, there is no reason why they are the way they are.

This argument for locality (unlike those we saw earlier) is not circular. It relies upon an idea pretty far removed from locality: that there is an explanation (as yet unknown) of the details of these force laws. So we have made *some* progress. But we are not done yet. We must still find some reason to believe that the details of these force laws have explanations. We might suggest that *every* fact must have an explanation: Nothing happens without a reason – something making *it* happen rather than any alternative. This is sometimes called the "principle of sufficient reason" and was famously advocated by Gottfried Leibniz (German philosopher, scientist, and mathematician, 1646–1716).

If our argument for locality requires the principle of sufficient reason, then we must investigate that principle's plausibility. Must every fact have an explanation? One fact is explained by others. If those other facts have explanations, then they are explained by still other facts. And so on. How, then, could *every* fact have an explanation?

One possibility is that this sequence of explanations never ends; for every fact, there is another, "deeper" fact that explains it, and so forth infinitely. But then science could never, even in principle, achieve its goal of discovering the most basic explanatory principles, since no principles really are the "most basic." A "theory of everything," giving the reasons for all facts for which there are reasons, must necessarily elude us (although for any fact having an explanation, science could in principle discover its explanation). This would come as a disappointment for someone like Steven Weinberg (American physicist, b. 1933, winner of the 1979 Nobel Prize), who in his book *Dreams of a Final Theory* wrote of physics as aiming to discover the "starting point, to which all explanations may be traced" (1993: 6). At the very least, Feynman says, "it [would] become boring that there are so many levels one underneath the other" (1965: 172).

It would also be a bit strange. Rather than nature being governed by some fundamental rule-book, every rule would instead be merely the upshot of some more basic rules, which themselves would fail to capture the game's fundamentals since they, in turn, would merely reflect some still more basic rules, and so on indefinitely. (No actual game, such as tennis or checkers, works this way.) There would be no genuine "*first* principles" from which every constant of nature (such as the speed of light) logically follows. Many would agree with Feynman that there is a bottom level: "This thing cannot keep on going so that we are always going to discover more and more new laws" (1965: 172).

Is there any other way for every fact to have an explanation (thereby supporting our argument against action at a distance)? You might suggest that the sequence of explanations eventually ends with facts that somehow explain themselves. But how could a fact explain itself? The only plausible suggestion I know is that certain truths *that could not have been false* (such as that all squares have four corners) count as explaining themselves.[1] However, as I suggested in chapter 3, the various force laws presumably *could* have been false, as in certain science-fiction stories. They are not logically necessary. So they cannot follow solely from truths that could not have been false.

Alternatively, you might suggest that every fact would have an explanation if the sequence of explanations ran in a circle: fact B helps to explain fact A, and fact C helps to explain B, and D helps to explain C, etc., and eventually, we reach some fact that A helps to explain. But if explanations could run in circles like that, then A could be used to help explain itself. As we just saw, this is not plausible when the facts in question are not logically necessary.

I see no alternative but to deny that every fact has an explanation. On this view, some facts are unexplained explainers. The point is not that *we do not know* why these facts hold, but rather that *there is no reason* why. These facts are *brute*; they are just the way things happen to be, for no reason at all. Brute facts are facts that have no explanation and yet could have failed to hold.[2]

By an argument like this, we could know that there are brute facts even if we do not know which facts these are, since any candidates could someday be discovered to have explanations. Although the foregoing is hardly a knockdown argument against the principle of sufficient reason, it is strong enough to prevent us from taking the principle of sufficient reason for granted in an argument for locality.

That there may well be brute facts affects an argument we saw in chapter 2 aiming to show that the electric potential is unreal. Maxwell's equations say that nothing depends on the electric potential's absolute value; it is a dangler. The best explanation of this fact is arguably that the potential is unreal – that there is no absolute value of the electric potential. But another possibility, difficult to discount, is that Maxwell's equations are brute facts, and so there is no explanation of why the electric potential's absolute value makes no difference to anything else. It just does; for no reason, it exists as a dangler.

The conclusion that some facts are brute might meet with the following objection. Suppose that a rainbow appears in the sky. Why does this happen? Because water droplets are suspended in the air, sunlight is shining through them, and various laws of refraction and reflection govern the light's interaction with the droplets. Why are there some water droplets in the air, rather than none there? Because it rained earlier in the day and so on. Why did it rain? Ultimately, we reach facts that have no explanations. For instance, consider why the laws of refraction and reflection hold. Presumably, the answer involves Maxwell's equations, and perhaps these are brute facts. But if our explanation of the rainbow's occurrence appeals to facts that have no explanation themselves, then (the objection runs) we have not really explained why the rainbow appeared. We have shown that the rainbow occurred because certain other facts held, but we have not explained why those other facts held, and so we have not yet accounted for the rainbow's appearing. If all so-called explanations rest ultimately on facts having no explanation, then none of these "explanations" really succeeds.

This objection is based on an incorrect picture of how scientific explanation operates. We do not need to know *why* there were water droplets in the air in order to use that fact to explain why the rainbow occurred. We need to know only *that* there were water droplets in the air (Hempel 1973). An explanation can be *complete* even if the facts to which it appeals have no explanations. Of course, if we asked why a rainbow appeared and were told that there were water droplets in the air, this answer *might* lead us to wonder why there were water droplets in the air. But this is a different question from the one we originally asked. That the answer to the first question led us to ask the second question obviously does not mean that a complete answer to the first question must include an answer to the second.

Why is it that we can use A to account for B even if we don't know why A is the case, and even if there is no reason why A is the case? A scientific explanation answers a certain question: Why B? That question is asked within some context determining what would count as a satisfactory answer. For example, a physician who wants to know why a child has hives will be content to discover that the child was stung by a bee and has an allergy to bee stings. For the physician's purposes, it is not necessary to find out why the bee stung the patient. The physician is likewise uninterested in an explanation of the case of hives that begins all of the way back before the bee stung the patient. To explain the hives, it's enough for the physician to know that the patient was stung. On the other hand, the patient's parents might have different

---

Box 4.1
Brute facts and satisfying explanations

That the fundamental laws of nature, lying at the bottom of all explanatory chains, are *brute* facts does not mean that they cannot constitute a satisfying stopping-point for scientific explanation. Einstein (1949: 63) and Weinberg (1993: 234–7) have speculated that the fundamental laws will turn out to be a satisfying place for scientific explanations to end because the fundamental laws could not have been *slightly* different. In other words, even though the fundamental laws could still have been *vastly* different from the way they actually are, perhaps any slight change in their form, or in the values of the physical constants, would logically entail contradictions (such as probabilities less than zero) or ever greater departures in the fundamental laws (so that the change cannot remain slight).

---

concerns in asking why the hives occurred. They might want to know why their child was stung in the first place – say, that he was throwing rocks at a bee hive, exciting the bees. Their interests do not allow them to rest content with an explanation that begins with the bee sting. On the other hand, they are not interested in tracing the causal antecedents of the bee sting all the way back to details of the Big Bang.

Thus, to know why a rainbow appeared, we do not (in a typical context) have to know why the laws of refraction and reflection hold or why there were water droplets in the air. Similarly, for the laws of electricity and magnetism to have explanations, their explainers do not have to have explanations, and so on indefinitely. The ultimate explainers of the laws of electricity and magnetism could be brute (see box 4.1) – and could be laws of contact action. But is there any reason to expect this to be the case?

3   Which Facts are Brute?

Let us try to refashion our argument for locality so that it no longer relies on the dubious principle of sufficient reason. Perhaps we can admit the existence of brute facts and still argue for locality. The argument I have just given, concluding that there must be some brute facts, does not tell us *which* facts are brute. This is a crucial point. Consider the fact that the electrostatic force diminishes with the square of the distance, or the fact that the speed governing the time delay is about 300,000,000 m/s. Are these the sorts of facts that we would expect to be brute (assuming that there are brute facts)? If not, then we have here an argument for locality (insofar as we believe that these facts would be brute under retarded action at a distance but not under locality).

There are many episodes in the history of science in which one theory was judged to be more plausible than another not because it was more accurate with regard to current observations or because it took *fewer* facts as brute, but rather because it was thought to portray the *right sorts* of facts as brute. That is, one theory was judged more plausible than others in virtue of its offering explanations of those facts that were thought more likely to have explanations than to be brute, and its considering to be brute those facts that were thought more likely to be brute than to have explanations. For instance, Johannes Kepler (German astronomer, 1571–1630) used this sort of argument to advocate the model of the solar system proposed by Nicolaus Copernicus (Polish astronomer, 1473–1543). According to Copernicus, the Earth and the other planets orbit the Sun, contrary to the model proposed by ancient astronomers such as Claudius Ptolemy (who lived in Alexandria between about the years 100 and 170), according to which the Earth is orbited by the Sun, Moon, and other planets. Consider a "superior" planet – that is, a planet that takes longer than one Earth year to complete a circuit through the zodiac. Mars, Jupiter, and Saturn were the superior planets known. Since ancient times, it had been noticed that a superior planet is at "opposition" (that is, rises at sunset and sets at sunrise as seen from Earth, so that it is "opposite" to the Sun) always and only when it is brightest (which was understood to be when it is nearest to Earth – that is, at its "perigee"). On Ptolemy's Earth-centered model, this correlation between opposition and perigee was explained by fine-tuning the rates at which the planets and Sun move on their respective circular paths. The relation between these rates that is needed to explain the opposition–perigee correlation is, on this model, a brute fact (see figure 4.1).

On the other hand, Kepler noted that on Copernicus's model, the correlation between a superior planet's opposition and perigee is explained *automatically* just by the geometry involved in having the Earth orbiting the Sun inside the orbit of the superior planet (see figure 4.2).

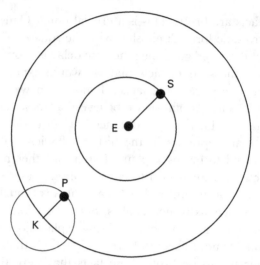

Figure 4.1 The relation between the Sun (S) and a superior planet (P) at opposition to the Earth (E) according to an Earth-centered theory. The Sun is orbiting the Earth while the planet is orbiting on a small circle (the "epicycle"), the center (K) of which is orbiting the Earth on a large circle (the "deferent"). All orbits are counterclockwise. As long as the planet orbits in step with the Sun, so that line KP always remains parallel to line ES, the planet will reach its closest point to the Earth (the "bottom" of its epicycle) exactly when the Sun is at the opposite side of the Earth from the planet.

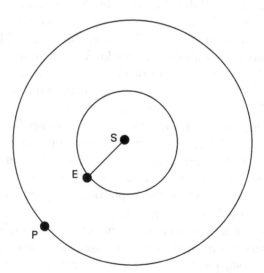

Figure 4.2 Opposition according to a Sun-centered theory. It automatically occurs when the planet (P) is nearest to the Earth (E).

Opposition requires that the Earth be directly between the planet and the Sun, since the planet is then on the opposite side of the Earth from the Sun (so sunrises coincide with planet-sets). On the Copernican model, then, the planet is nearest the Earth when it is at opposition. (Try moving the planet along its orbit in figure 4.2 while leaving the Earth exactly where it is. You can't help but increase the planet's distance from the Earth.) No fine-tuning of orbital speeds is needed to achieve this result; it follows automatically from the geometry.

Kepler saw this as evidence for Copernicus's model, as he explained in 1596:

> My confidence [in Copernicus's theory] was first established by the magnificent agreement of everything that is observed in the heavens with Copernicus's theories. . . . However, what is far more important than that, for the things at which from others we learn to wonder, only Copernicus magnificently gives the explanation, and removes the cause of wonder. . . . [R]easons are supplied for a great many . . . matters for which Ptolemy for all his many motions could give no reason. . . . [T]he ancients rightly wondered why the three superior planets are always in opposition to the sun when they are at the bottom of their epicycles [that is, at perigee]. . . . In Copernicus's theory the reason is easily supplied. For it is not Mars on one epicycle but the Earth on its own circle which causes this variation.   (1981: 75–6, 81; cf. Jardine 1984: 141, 145)

Of course, Kepler is overreaching a bit in saying that the Ptolemaic model fails to explain why oppositions coincide with perigees. The explanation, according to Ptolemy, is that there is an exquisite coordination between the orbital rates of the Sun and superior planets. Admittedly, this coordination itself has no explanation, but in the previous section, we saw some reason to expect there to be brute facts. Kepler's argument, then, should be that this coordination between the orbital velocities of the Sun and superior planets would have to have an explanation; such a remarkable relationship, we tend to feel, would not hold without a reason. Thus, unlike Ptolemy's theory, Copernicus's theory explains the facts that intuitively should have explanations and portrays as brute the facts that would intuitively likely be brute.

Let us look at another, more recent example. The ratio of the universe's gravitational potential energy (reflecting gravity's tendency to slow down the universe's expansion from the Big Bang) to its kinetic energy of expansion (reflecting the rate at which bits are flying apart) determines the universe's geometry and its fate: whether it is going to expand forever or collapse in a Big Crunch. This ratio, Ω (omega, the final letter of the Greek alphabet, as befits the factor determining the universe's final state), is difficult to ascertain, but observational data lead cosmologists to be fairly sure that it lies between 0.1 and 10.0. In other words, the kinetic energy and the gravitational potential energy are nearly equal. (In cosmology, two numbers within a couple of powers of ten are nearly equal, considering that cosmologists routinely discuss differences of 119 orders of magnitude between the universe's density now and 10^{-43} seconds after the Big Bang.) That today Ω is approximately equal to 1 ($\Omega \approx 1$) entails that there will be no Big Crunch and that geometrically, the universe resembles an infinite flat surface.

On the *original* Big Bang model, Ω's current value results directly from a brute fact about the initial conditions of the universe: the value that Ω just happened to take right after the Big Bang. The original Big Bang model says that $\Omega \approx 1$ today, 10 billion years after the Big Bang, if and only if at 10^{-43} seconds after the Big Bang, Ω was between $1 - 10^{-59}$ and $1 + 10^{-59}$ (Lightman and Brawer 1990: 23). But many cosmologists regard the new *inflationary* Big Bang model as giving a better explanation of Ω's current value. The inflationary model does not directly attribute that value to a brute fact. According to this model, there was a brief period of very rapid cosmic expansion before the universe was about 10^{-32} seconds old. The process hypothesized as powering this expansion turns out to dictate that no matter what Ω's initial value had been, Ω would have reached a value close to 1 (in any observable patch of the universe) by the end of the inflationary period.

One of the arguments that cosmologists give for the inflationary model is apparently that Ω's current value should have an explanation rather than be the direct consequence of a brute initial condition. Responding to the original Big Bang model's appeal to Ω's initial value, the distinguished cosmologist D. W. Sciama says:

> Such "fine tuning" looks unnatural, and suggests that either the density is indeed exactly critical [that is, $\Omega = 1$ now], for a reason still to be discovered, or that some non-standard mechanism intervened to drive the present value of the density close to the critical value.... [T]he inflationary universe claims to provide such a mechanism. (1983: 495)

But what is meant by the supposed initial conditions on the original Big Bang model looking unnatural?

The argument should not be that *any* direct appeal to brute facts is prohibited, since in the previous section, we saw some reason to think that there must be brute facts. We feel pretty content to explain certain facts by straightforwardly appealing to brute initial conditions. For instance, consider the remarkable fact that the Sun's angular size in Earth's sky is almost exactly the same as the Moon's. That is one of the reasons why solar eclipses are so spectacular: the Moon fits nearly exactly over the Sun's disk, so that only the Sun's outer atmosphere peeks out from behind it. On the best current model of the solar system, the Sun's and Moon's sizes in Earth's sky are freely adjustable parameters. So what are the chances that they would turn out to be so nearly equal? Virtually zero, yet astronomers do not consider it a defect in the current model that it must hypothesize this coincidence in order to match our observations. Astronomers do not seek some deeper explanation of the equality between the Sun's and Moon's angular sizes in Earth's sky. In contrast, many astronomers currently believe that the rough equality between the two factors in Ω's initial value demands a deeper explanation than that these two factors are just brute facts. What evidence justifies these beliefs?

The argument for locality that we have been examining boils down to this question. Consider the fact that the speed governing the electromagnetic time delay is 300,000,000 m/s. Is this like the fact that Ω's initial value was very close to 1 in that neither fact is plausibly brute? If so, then (insofar as we believe that this speed would be brute under retarded action at a distance but not under locality) we would have an argument for locality. On the other hand, is the fact about the electromagnetic time delay like the fact that the Sun's and Moon's angular sizes are about equal – in that both facts are plausibly the direct results of brute facts? If so, then we would have here no argument for locality.

This question is difficult, and so any argument here for locality must at best remain inconclusive. But there are some significant and perhaps relevant differences between the electromagnetic time delay and Ω's initial value. To begin with, for Ω's initial value to be about 1 is for the energy powering the universe's expansion to be about equal to the energy slowing that expansion. Experience suggests that such a near balance *between opposing influences* tends not to be a coincidence; we have more often found a common cause for two such quantities when they are about equal than when their values fall in just any old ratio. Although the Sun's and Moon's angular sizes in Earth's sky are about equal, they are not *opposing influences*. Likewise, the speed governing the electromagnetic time delay does not concern opposing influences.

But this cannot be the whole story. Consider Ptolemy's relation between the orbital velocities of the superior planets and the Sun. This relation is not in any recognizable sense a near equality of opposing influences. Yet Kepler apparently thought it too remarkable to hold for no reason.

There is something special in the Big Bang theory about Ω being set initially at 1 rather than at any other value. According to the original Big Bang theory, $\Omega \approx 1$ today only if at 10^{-43} seconds after the Big Bang, Ω was between $1 - 10^{-59}$ and $1 + 10^{-59}$. That is because according to the Big Bang theory (original or inflationary), $\Omega = 1$ is an unstable equilibrium (like a pencil balanced perfectly on its point). In other words, once $\Omega = 1$, it will remain there, but if Ω is set initially even slightly different from 1, then as time passes, it will diverge increasingly from 1. Only at $\Omega = 1$ is there a point of equilibrium.

There is a further respect in which the Big Bang theory sets apart $\Omega = 1$ as a special situation compared to Ω being any other value – a further respect in which $\Omega = 1$ is "privileged" by the Big Bang theory. It turns out that $\Omega = 1$ is a critical point in determining the geometry and fate of the universe. If $\Omega > 1$, then the universe is "closed": it will end in a Big Crunch. If $\Omega < 1$, then the universe is "open": it will expand forever. The only value of Ω that has this sort of privileged status in the theory is $\Omega = 1$.

Doubtless for any value that Ω might have, we can find *something* distinctive about it, as the Indian mathematician Srinivasa Ramanujan (1887–1920) teaches us in a famous episode recalled by the English mathematician G. H. Hardy (1877–1947):

> I remember once going to see him when he was lying ill at Putney. I had ridden in taxi-cab No. 1729 and remarked that the number seemed to me rather a dull one, and that I hoped it was not an unfavourable omen. "No," he replied, "it is a very interesting number; it is the smallest number expressible as the sum of two cubes in two different ways. [$1^3 + 12^3 = 9^3 + 10^3 = 1,729$.]" (Mackay 1977: 127)

It would be difficult to say what it takes in general for a certain value of some parameter to have a "privileged status" in a given theory – to be special "from the theory's own viewpoint." But perhaps even without a general account, we can recognize that the Big Bang theory privileges only one setting for Ω – namely, $\Omega = 1$. The fact that Ω's initial value, according to the original Big Bang theory, would have to be coincidentally very near to the only value privileged by that theory seems to me a good reason for being suspicious about that theory. It is a reason, in other words, for characterizing that theory's initial setting of $\Omega \approx 1$ as unnatural fine-tuning.

It is unclear to me whether this argument is based on our scientific experience, like the argument regarding a balance between opposing influences, or whether its ground is independent of the way that we have discovered the universe to be. But in any case, the same argument does not work against regarding as brute the fact that 300,000,000 m/s is the approximate speed governing the electromagnetic time delay. (However, this argument might work against regarding as brute the inverse-square character of gravitational forces, since an exponent of −2 is privileged: it alone permits stable orbits.)

We have arrived, then, at an impasse: action at a distance does not quickly lead to any obviously unsatisfactory consequences for scientific explanation. Even if we believe that various facts regarding electromagnetic forces would be unexplained if these forces operated by action at a distance but would be explained by the details of whatever local process were at work, this consideration does not obviously favor locality. For that matter, it is not certain that the details of a local process would indeed explain the speed governing the electromagnetic time delay, for instance. The nineteenth-century project of giving physical models of the medium conveying electromagnetic influences turned out to be difficult to reconcile with Einstein's special theory of relativity. However, I have been disregarding this point in order to show that the argument for locality is inconclusive even granting its assumptions about the explanations that a local account would supply. In the next chapter, we will look elsewhere for arguments regarding locality.

Discussion Questions

You might think about . . .

1   The fine structure constant α is a "pure number" (a quantity independent of the units being used) defined as $e^2/\hbar c$, where e is the fundamental charge (the electron's charge), $\hbar$ is Planck's constant divided by 2π, and c is the speed of light. One way to think about α is

as characterizing (independent of any choice of units) the strength of the electromagnetic force. It has sometimes been argued that the physical constants in α could not have been much different and still permitted living things to exist (Leslie 1989). (For example, had *e* been slightly different, then the stable isotope of carbon could not have formed.) So can the values of these constants be explained by the fact that we exist? (This idea is sometimes called "the anthropic principle.") After all, we would not have been here to know it if these constants had taken on much different values; the only values that living things could have encountered are values in an exceedingly narrow range around the actual ones. On the other hand, is it one thing to explain why those values hold, and quite another thing to explain why we observe those values to hold? Would it help to hypothesize that "there is not one universe but a whole infinite set of universes with all possible initial conditions" (Collins and Hawking 1973: 319)?

2   It might be argued that a goal of science is somehow to eliminate any unexplained explainers or, at least, to minimize them. As evidence for this view, it might be suggested that in any scientific explanation, the explainers must themselves be less in need of explanation than the fact being explained. As one advocate of this view has written:

> For instance, "Peter is falling past the window in the third floor, because one second ago he was falling past the window in the fifth floor" . . . is *not* adequate as an explanation, because the cause is here just as much in need of explanation as the effect.   (Schurz 1999: 97)

Do you agree? You might consider either of the following lines of thought – or pursue one of your own.

(a)   To explain why the dinosaurs, ammonites, and many other animal groups became extinct about 65 million years ago, scientists say that a rocky body about the size of Manhattan collided with Earth. (The collision produced the Chicxulub crater underneath Mexico's Yucatan peninsula.) Is this an explanation where the fact doing the explaining stands less in need of explanation than does the fact being explained? Can you think of better examples?

(b)   Explanations often involve explaining why some fact held *rather than* certain alternatives. For instance, in explaining why Smith died of lung cancer, we might be explaining why Smith *died* of lung cancer rather than survived the illness, or why Smith died *of lung cancer* rather than failed to develop lung cancer, or perhaps why *Smith* rather than Jones died of lung cancer (considering that both were heavy smokers). Many jokes depend on the alternatives implicit in why-questions. For example, when my daughter, Rebecca, was six years old, she told me this joke: "Why did the dinosaur cross the road? Because the chicken had not evolved yet!" The joke is based on the fact that the listener interprets the question as "Why did the dinosaur cross the road rather than fail to cross it?" whereas the answer explains why the dinosaur rather than the chicken crossed the road. In asking why Peter is falling past the third-floor window, what would the contrast typically be – falling past the third-floor window *rather than what*? Would "Because one second ago he was falling past the fifth-floor window" answer this question?

3   Robert Boyle (English natural philosopher, 1627–91) discusses where the sequence of explainers comes to an end. He seems to think that the basic features of nature that explain all the rest can themselves be explained theologically: God's existence requires no explanation, since God could not have failed to exist, and God's nature requires that God create a universe with exactly those basic features possessed by the actual universe. Boyle says:

> [T]here are diverse effects in nature, of which, though the immediate cause may be plausibly assigned, yet if we further enquire into the causes of those causes, and desist not from ascending in the scale of causes till we are arrived at the top of it, we shall perhaps find the more catholic and primary causes of things to be either certain prim- itive, general and fixed laws of nature, or rules of action and passion among the parcels of the universal matter; or else the shape, size, motion, and other primary affections of the smallest parts of matter . . . – of all which it will be difficult to give a satisfactory account without acknowledging an intelligent Author or Disposer of things. (1991: 156–7)

What do you think of Boyle's final remark?

4   Does Ptolemy's relation between the orbital velocities of the super- ior planets and the Sun have a "privileged status" among the various relations they might have had in Ptolemy's model? Why or why not?

Notes

1 Nozick (1981: 118–21) entertains another, somewhat more outlandish means whereby a fact might explain itself.

2 It might be suggested that every universe involving no contradiction really exists ("in parallel" with all of the others). Each of these universes has its own laws and initial conditions. These give the universe its identity, individuating it from the other universes. On this view, consider the fact that this universe (the one we actually inhabit) has such and such laws and initial conditions. This fact would be an unexplained explainer but not brute, since this universe necessarily has those features; if it had different ones, it would be one of the other universes. However, this suggestion still leaves us with a brute fact: that every non-contradictory universe exists. Nozick (1981: 131–7) considers whether this fact could explain itself.

5
Fields, Energy, and Momentum

1 Introduction

Suppose the electric field really exists and an electrically charged point body at location A interacts with another electrically charged point body at location B some distance away. (Assume for simplicity that surrounding these bodies, there is a vacuum.) Then the effect at A (a force) has a complete set of causes at A: the charge and electric field there. This interaction obviously obeys spatiotemporal locality. The field at A was affected, in turn, by the body at B. This causal relation also obeys spatiotemporal locality. As I explained in chapter 2, a charged body's arrival (at time t_0) at B causes an event that moves continuously outward from B, like one of the ripples that is produced when a stone is tossed into a pond. The speed at which this effect event moves equals the speed of light (c). So at a time t shortly after t_0, the effect event is located at exactly the points in space at a distance from B equal to $c(t - t_0)$; the distance that the effect event has traveled equals the event's speed multiplied by the time elapsed. The points at which the effect is located at a given moment form a spherical surface centered on B.

At a given location, the effect consists of some contribution to the electric field there – namely, the addition to the field of a component that reflects the charged body's arrival at B. Before the effect can travel to a location at some distance from B, the effect must first come into existence local to B. In other words, the effect begins (at t_0) at the same moment as its cause occurs (the charged body's arrival at B) and at that moment, the effect is not separated from its cause by any region of space. Therefore, this effect is spatiotemporally local to its cause (according to the definition of "spatiotemporal locality" in chapter 1).

The charged body's arrival at B affects the electric field at A, but this effect does not take place instantly upon the body's arrival at B. Rather, there is a delay: it takes some time for the ripple on the pond to cross the space between B and A. In particular, if r is the distance between B and A, then it takes an interval of time lasting for r/c. The effect at A is said to be "retarded" – that is, time-delayed.

So if fields are real, the electric and magnetic interactions between bodies obey spatiotemporal locality. On the other hand, if fields are *not* real, then there is *retarded action at a distance*. Just as on the field picture, the charged body's arrival at B affects the force felt by the charged body at A, and this effect is delayed by a time interval of r/c. But in contrast to the field picture, there are no complete causes nearer in space or time to the effect. In other words, between the effect and some members of its nearest complete cause, there is a gap in space (between A and B) and a gap in time (of r/c). Both spatial and temporal locality are violated. The time delay is not explained by the need for some cause to make its way across the space between B and A; the retardation is presumably a brute fact, as we saw in chapter 4.

Because this action at a distance is *retarded*, it turns out to violate the laws of energy and momentum conservation. This objection to retarded action at a distance forms the main subject of this chapter. It supplies a strong argument for the reality of fields, and hence for spatiotemporal locality. But this line of argument also raises a variety of thorny problems involving the nature of energy and momentum. We will have to work through these problems in this and subsequent chapters.

2 The Argument from Conserved Quantities

In classical physics, energy and momentum are conserved in any closed system. (That is, neither the system's energy nor its momentum changes over time.) Let us think about these two quantities very simply. A body's momentum is a vector quantity. That is to say, it has an amount and a direction. In classical physics, the amount (the "magnitude") is equal to the product of the body's mass m and its speed v. The direction of a body's momentum vector is identical to the direction in which the body is moving. That is to say, a body's momentum is mv, where the vector v is the body's "velocity." (The magnitude of v is the body's speed; v points in the direction in which the body is moving.)

Energy is a scalar, not a vector; it has a magnitude (usually measured in "ergs" or "joules") but no direction. A body's energy is equal to its "kinetic energy" plus its "potential energy." For now, a brief sketch of these concepts will be enough. A body's kinetic energy is its energy of motion, defined as half the product of its mass with the square of its speed: $(\frac{1}{2})\, mv^2$. Its potential energy is its energy of position and is greater insofar as the body is farther away from the location to which it "wants to go." For instance, suppose the bodies at A and B both possess charge q, so they repel each other. They "want to go" as far away from each other as possible. So the pair's electric potential energy is zero when the two bodies are infinitely far apart. The closer they are pushed together, the greater their electric potential energy. The electric potential energy of the pair is q^2/r, as we shall see later in the chapter.

This sketch will be enough to reveal how both energy and momentum conservation are violated if the bodies at A and B interact by retarded action at a distance. Suppose that for a long period of time ending shortly before t_0, one body is held at rest at A, whereas the other body is held at rest at some location farther away from A than B is. Because they have the same electric charge, they repel each other from that distance (call it R), but they are held in place by some other forces; perhaps two people hold them so that they do not move. Then shortly before t_0, the body not at A is moved nearer to A; in particular, it is moved to B, at a distance r from A (figure 5.1). At t_0, it is released at B so that it is momentarily at rest. At the same instant, the body at

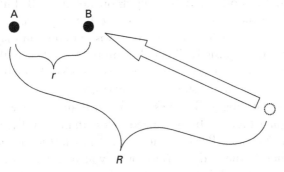

Figure 5.1 Body B is moved nearer to body A, their separation decreasing from R to r.

A is released at rest. At t_0, then, the two bodies are at rest, separated by distance r, and each is free to move. Once the two bodies are released, the system consisting of the two bodies is closed. (Before they were released, the two bodies did not form a closed system because each was interacting with the hands holding it. Assume that the two bodies, once released, are far enough away from any other body that we can treat the system consisting of the two bodies as isolated from any other body.) Since the two bodies form a closed system, their total energy must remain constant as they interact, and so must their total momentum.

What electric force does the body at B feel at t_0? We have assumed that the force results from retarded action at a distance. So the repulsion felt at t_0 by the body at B reflects the location of the other body at some earlier moment. However, even at that "retarded time," the other body was still at A; I stipulated that it has been there for a long time, so there has been enough time for its presence at A to make itself felt at B, a distance r away. Therefore, by Coulomb's law (equation 2.1), the body at B feels a force repelling it from A equal to q^2/r^2. Now what force does the body at A feel at t_0? Again, because the influence of the body at B is retarded, the repulsion felt by the body at A reflects the location of the other body at some earlier moment. The other body has only just arrived at B, and it takes a time interval of r/c for "news" of its arrival at B to reach A. So at t_0, that "news" has not yet reached A. Just as the light we see at night from stars does not directly indicate what those stars are like today, but what they were like thousands of years ago, so the electric force felt by the body at A is caused by the other body as it was sometime earlier − when it was at a distance R away. What the body at A feels is a force equal to q^2/R^2, a weaker force than the body at B feels (since $r < R$). Furthermore, the force felt by the body at A is a repulsion not from B, but from the place at which the other body was located before it was moved to B. At t_0, then, the force felt by the body at A is not equal in magnitude to (and may not be oppositely directed from) the force felt by the body at B. Action does not equal reaction (contrary to Newton's third "law") because the action at a distance is retarded.

The body traveling to B moves *more slowly* than the "news" traveling to A (assuming that the body's speed is less than c). Why, then, does the body arrive first? Because the "news" has farther to go; it must reach A, whereas the body ends its journey at B. What does it take for the body to arrive at B before any "news" of its journey to B arrives at A? Suppose that the body's journey to B takes 500 seconds. In that time, "news" of the journey ("traveling" at c) can reach a distance about equal to Earth's distance from the Sun. So in order for "news" of the journey's start to arrive at A only after the journey has been completed, A's separation R from the location at which the journey began must exceed Earth's separation from the Sun. But if the body journeys in a straight line at 100 kilometers (60 miles) per hour, then it can cover in 500 seconds only about 14 kilometers, which is a very small fraction of its original distance R from A (equal at least to Earth's distance from the Sun). Therefore, r's difference from R will be a very small fraction of R, and so the difference between the force felt by the body at B (q^2/r^2) and the force felt at the same moment by the body at A (q^2/R^2) will be negligible for practical purposes. Action will very nearly equal reaction. So in our ordinary lives, we notice no violations of Newton's third law.

Nevertheless, there is a slight inequality between action and reaction. Consequently, the total momentum and total energy of the two bodies in the moments following t_0 fail to equal the system's total momentum and total energy at t_0. This is the objection to retarded action at a distance: it conflicts with energy conservation and momentum conservation (box 5.1 gives the details).

A conflict between electromagnetic theory and the conservation laws can be avoided if the interaction between the bodies at A and B does not operate by retarded action at a distance. Instead, suppose each body interacts with the electric field at its location, in accordance with spatiotemporal locality. If the field carries energy and momentum, then each body withdraws some energy and momentum from the field. Whatever quantity the bodies gain is exactly the quantity lost by the field. Consequently, energy and momentum are conserved. When we thought of these interactions as retarded action at a distance, we considered the two bodies (after t_0) as forming a closed system. But if the electric field is real, then the two bodies fail to form a closed system; they are exchanging energy and momentum with the field. An open system does not have to obey energy and momentum conservation, accounting for the bodies' failure to do so. The system consisting of the two bodies *together with the field* is closed. The energy of the field, added to the energies of the bodies, remains constant throughout the interval, and likewise for momentum.[2]

To elaborate this picture, we need expressions from Maxwell's electromagnetic field theory specifying how much energy and momentum the field possesses. I will investigate these expressions later in the

Box 5.1
How energy and momentum conservation are violated
by retarded action

Consider a very brief interval of time δt ("delta t," a very small number) following t_0. Because the interval is so short, we can use the approximation that the body starting the interval at A feels the same force throughout the interval: q^2/R^2 directed away from the location of the other body before it was moved to B. This force, multiplied by the interval δt during which it is felt, gives (according to Newton's second law of motion) the change in the body's momentum over the course of the interval. (Newton's second law says that the rate of change in a body's momentum equals the force on the body.) By the same reasoning, the other body's momentum changes by $\delta t \, q^2/r^2$ directed away from A. So the two momentum changes are not equal and opposite; they have different magnitudes and may not even be directed oppositely. So the system's final momentum is unequal to its initial momentum. Retarded action at a distance violates momentum conservation.

Now consider energy conservation. When released at t_0, both particles are at rest, so neither has any kinetic energy. The pair's potential energy is q^2/r. So:

$$\text{The pair's total initial energy} = q^2/r. \qquad (5.1)$$

Compare this to their total energy at the end of the interval. We have just seen that after δt has passed, the momentum mv of the body starting the interval at A is $\delta t \, q^2/R^2$. By dividing its momentum by its mass m, we determine its speed v at the end of the interval to be

$$\delta t \, q^2/mR^2. \qquad (5.2)$$

By taking $(1/2)mv^2$, we determine its kinetic energy: $(1/2)m(\delta t \, q^2/mR^2)^2$, which simplifies to $q^4(\delta t)^2/2mR^4$. By analogous reasoning, the other body's kinetic energy at the end of the interval (assuming, for simplicity, that its mass is also m) is $q^4(\delta t)^2/2mr^4$. So at the end of the interval, the system's total kinetic energy is

$$q^4(\delta t)^2/2mR^4 + q^4(\delta t)^2/2mr^4,$$

which is

$$[q^4(\delta t)^2/2m][1/R^4 + 1/r^4]. \qquad (5.3)$$

Now we need to find the pair's potential energy at the end of the interval. It is q^2 divided by their separation at the end of the interval. Again approximating each body as feeling a constant force (and so undergoing uniform acceleration) during the interval, each body moves a distance given by the time elapsed (δt) multiplied by its average speed during the interval, which is half of its speed at the end of the interval. In (5.2), we had the speed at the end of the interval of the body starting the interval at A. So that body moves a distance $(\delta t)^2\, q^2/2mR^2$, while the other body moves a distance $(\delta t)^2 q^2/2mr^2$. For simplicity, let's assume that a straight line could be drawn from A through B to the location of the body before it was moved to B. Then during the interval, the two bodies move in opposite directions.[1] So having started the interval separated by r, their separation at the end of the interval is

$$r + (\delta t)^2 q^2/2mR^2 + (\delta t)^2 q^2/2mr^2. \tag{5.4}$$

The pair's potential energy at the end of the interval is therefore q^2 divided by expression (5.4):

$$q^2/[r + (\delta t)^2 q^2/2mR^2 + (\delta t)^2 q^2/2mr^2]. \tag{5.5}$$

Let's simplify this monster; it is q^2/r times

$$1/[1 + (\delta t)^2 q^2/2mrR^2 + (\delta t)^2 q^2/2mr^3]. \tag{5.6}$$

Now we use the approximation that for any small quantity δ,

$$1/[1 + \delta] = 1 - \delta.$$

Using this approximation to simplify (5.6), the pair's final potential energy is q^2/r times

$$1 - (\delta t)^2 q^2/2mrR^2 - (\delta t)^2 q^2/2mr^3,$$

which (after pulling out common terms) is q^2/r times

$$1 - (\delta t)^2 q^2/2m\ [1/rR^2 + 1/r^3],$$

which (after putting back the q^2/r) is

$$q^2/r - [q^4(\delta t)^2/2m]\ [1/r^2R^2 + 1/r^4]. \tag{5.7}$$

Now we're almost done. The system's total energy, kinetic (5.3) plus potential (5.7), at the end of the interval is therefore

$$[q^4(\delta t)^2/2m][1/R^4 + 1/r^4] + q^2/r - [q^4(\delta t)^2/2m] [1/r^2R^2 + 1/r^4].$$

Rearranging and combining terms, we get

$$q^2/r + [q^4(\delta t)^2/2m][1/R^4 + 1/r^4 - 1/r^2R^2 - 1/r^4].$$

A very welcome cancellation then gives us our final expression for the system's total energy at the end of the interval:

$$q^2/r + [q^4(\delta t)^2/2m][1/R^4 - 1/r^2R^2]. \tag{5.8}$$

This is nice. The first term is identical to (5.1): the system's total energy at the beginning of the interval. The second term is non-zero because r is unequal to R. So we have shown that the law of energy conservation is not obeyed by retarded action at a distance!

Let us verify that the culprit here is the *retarded* character of the action at a distance. Imagine that the electric interaction between the two bodies is instantaneous (that is, non-retarded). Because there is now no time delay, the body at A feels a force at t_0 (given by Coulomb's law) of q^2/r^2. Again approximating as constant the force on this body over the brief interval, the body's momentum changes over the interval (according to Newton's second law) by the force it feels multiplied by δt. Since the body began the interval with no momentum (it was at rest), its momentum mv at the end of the interval is $(\delta t)q^2/r^2$, and so its speed then is $(\delta t)q^2/mr^2$. (Would you like to pick up a pencil and take it from here? It's kinda fun . . .) Its kinetic energy $(1/2)mv^2$ at the end of the interval is therefore

$$(1/2)m[(\delta t)q^2/mr^2]^2. \tag{5.9}$$

The same reasoning applies to the other body, so at the end of the interval, their total kinetic energy is double that of the first body. Doubling (5.9), simplifying yields

$$q^4(\delta t)^2/mr^4. \tag{5.10}$$

Now let's find their potential energy at the end of the interval. As before, we reason that the distance each body travels in the interval is δt multiplied by its average speed during the interval, which is half of its final speed $(\delta t)q^2/mr^2$. So each body moves a distance $(\delta t)^2q^2/2mr^2$ away from the other. Their separation at the end of the interval is thus $r + (\delta t)^2q^2/mr^2$. So the potential energy of the pair of bodies at the end of the interval is

$$q^2/[r + (\delta t)^2q^2/mr^2].$$

This is q^2/r times

$$1/[1 + (\delta t)^2q^2/mr^3]. \tag{5.11}$$

We again use the approximation

$$1/[1 + \delta] = 1 - \delta$$

for any small quantity δ. Using this approximation to simplify (5.11), the pair's potential energy at the end of the interval is q^2/r times

$$1 - (\delta t)^2q^2/mr^3,$$

which (after putting back the q^2/r) is

$$q^2/r - (\delta t)^2q^4/mr^4. \tag{5.12}$$

But the second term of (5.12) is just the negative of (5.10) – the bodies' total kinetic energy at the end of the interval. So the sum of their kinetic and potential energies at the end of the interval is q^2/r, which is the same as (5.1): their total energy at the start of the interval!

Thus, energy conservation is upheld by instantaneous action at a distance. That is why classical mechanics, with its law of energy conservation, needed no field concept in order to incorporate Newtonian gravitational theory, according to which a change in the Sun's mass would instantaneously affect the gravitational force on Earth. It took electricity and magnetism to motivate field theory.

chapter. The moral of the story, so far, is that energy and momentum conservation suggest a powerful-looking argument for the field's reality, and hence for spatiotemporal locality.

3   Why Energy's Ontological Status Matters

How might a friend of retarded action at a distance reply to our argument from energy and momentum conservation to the reality of fields? She might reject the conservation laws that this argument takes for granted. But a less drastic response is available: she could reject the expressions we used for the momentum and energy of the two charged bodies. By using (in box 5.1) the standard expressions for the bodies' energies, we found (in expression 5.8) that the bodies' total energy at the end of the interval differed from their initial energy by $[q^4(\delta t)^2/2m]$ $[1/R^4 - 1/r^2R^2]$. A friend of retarded action at a distance might argue that the standard expressions for the bodies' energies must be mistaken, and that if we used the correct expressions, we would find that our original calculation of the bodies' energy at the end of the interval left something out: a term of precisely $[q^4(\delta t)^2/2m][1/r^2R^2 - 1/R^4]$! Thanks to this term, the bodies' total final energy would equal their total initial energy.

Where might this additional energy term come from? I mentioned that if we add the bodies' energies to the field's energy, as specified by a certain formula from Maxwell's theory (that I shall give later in this chapter), then we arrive at a conserved quantity. Could the friend of retarded action at a distance take Maxwell's formula and instead of interpreting it as the field's energy, simply use it in her calculations for the energy of the two bodies?

The expression for the system's energy would then include an additional term (from Maxwell's formula) – standing for what? On an action-at-a-distance interpretation, this term obviously cannot represent the field's energy. So it must stand for energy belonging to some bodies. Notice, however, that in our example of two charged bodies, this term in the expression for the system's energy at the *end* of the interval involves $[q^4(\delta t)^2/2m][1/r^2R^2 - 1/R^4]$ and so depends on R – the bodies' separation at an *earlier* moment, before the interval even began. Because the action at a distance is retarded, the additional energy at a given moment is a function not just of the bodies' properties *at that moment*, but also of their properties *at past moments*. Let's think about this.

Interpreted as standing for the field's energy, this term's dependence on R makes good sense. A charged body at a given moment makes a contribution to the field that propagates outward from that body's location at that moment. So the field's state at a given moment depends upon the past locations of various charged bodies, just as the pattern of concentric ripples in a pond at a given moment indicates the locations at which rocks in the recent past have plunged into the pond and how long ago these events occurred. So in our example involving two bodies, it is clear why the field's energy in the region surrounding those bodies would depend upon R, since some time ago, the bodies were separated by R.

But if fields are *not* real, then this additional energy term is somewhat peculiar in that it reflects the *history* of the system's bodies. Imagine two systems of bodies, each in isolation from the rest of the universe. The two systems consist of the same number of bodies. The bodies in one system are arranged in space relative to one another in exactly the same way as the bodies in the other system. Corresponding bodies in the two systems have the same mass, charge, velocity, and so forth. Imagine that the bodies in these two systems came to their current arrangements by different routes in space. In particular, suppose that each collection includes one body that moved to its current place quickly, having been located for a long period before at a great distance from the rest of the system, but that distance R differs for the two systems. Then the two systems *now* possess different energies, even though they consist only of bodies (no fields) and those bodies *now* are no different in their relative spatial arrangements and in all their intrinsic properties (except, obviously, their energies).

Of course, the friend of retarded action at a distance might be undeterred by this result. Admittedly, we are accustomed to thinking of a system's energy at a given moment as a function of its other properties at that same moment. For instance, a body's kinetic energy at a given moment is determined by its mass and velocity at that moment, and the electric potential energy of a pair of bodies at a given moment is fixed by their charges and separation at that moment. But perhaps this way of thinking about energy is mistaken.

How would we need to think about the system's energy? For each body in the system, the formula for the system's total energy includes a term representing its kinetic energy. For each pair of bodies in the system, the formula contains a term standing for the potential energy possessed by that pair. (In the next section, I will say more about *pairs* possessing potential energy.) Now how should the additional term in the formula (in our example: $[q^4(\delta t)^2/2m][1/r^2R^2 - 1/R^4]$) be interpreted from the perspective of retarded action at a distance? It must represent energy possessed by some body or bodies. Which ones? In our original example involving two bodies (at A and B), the energy must belong to that pair. But suppose we modify the example: instead of one body being moved quickly to location B, the body (a positively charged muon, let's say – an unstable subatomic particle) *decays* before it reaches B. The particles into which it is transformed (a positron and a neutrino) fly off in different directions. The body starting the interval at A feels a repulsive force during the interval. That force was directly caused some time earlier by the muon, even though the muon has decayed by the time the interval begins. (Likewise, tonight I may see light from a star that has long since exploded, the light having been caused by the star thousands of years ago.) The additional energy term during the interval reflects the electric charges of, and the earlier separation between, the muon and the body beginning the interval at A. But this additional energy cannot be possessed during the interval by the muon and the body starting the interval at A, since the muon has already decayed. Should we then ascribe the energy to the muon's decay products and the body starting the interval at A? On the one hand, we could imagine the muon passing its energy on to its decay products. On the other hand, the energy (as we shall see) did not really belong to the muon alone, but rather to the system consisting of the muon and the body starting the interval at A. This makes the "passing on" of the energy a bit harder to envision. In addition, if we ascribe the additional energy to the system consisting of the muon's decay products and the body starting the interval at A, then, strictly speaking, we cannot regard the additional energy as reflecting this system's history, since the muon's decay products were never separated by R from A; only the muon was. Accordingly, it may be most appropriate to ascribe the additional energy to the body at A and the muon's ghost! (Ohanian 1976: 80).

Perhaps the friend of retarded action at a distance should take a different approach: instead of looking desperately for *something* to possess this additional energy, she might deny that for each portion of the system's energy, there is some particular component of the system possessing it. On this view, the system's energy is not some stuff that the system possesses, distributed in some definite way among its various components. Rather, the system's energy is simply a mathematical quantity associated with the system, a quantity that remains constant as long as the system is closed. There is nothing, apart from the various quantities in the formula for the system's total energy, in virtue of which the system's energy is (say) 150 joules. There is no separate real property that the formula for energy represents. The system does not possess 150 units of some special stuff.

To understand this view, again consider the average family in the USA: facts about that "family" are nothing but complicated facts about the millions of "ordinary" families living in the USA. There is no real thing, the average family in the USA, that exists on a par with the "ordinary" families and whose characteristics make true certain statements such as "The average family in the USA has 2.1 children." Likewise, on the interpretation of "energy" that I am describing, the system's energy is not some entity or property that exists on a par with the system's bodies, their masses, charges, and so forth. "Energy" is just shorthand for a certain useful algebraic combination of various quantities characterizing the system. This combination is useful because its value remains constant over time. Energy conservation is a useful tool for figuring out what will happen because it does not require that we calculate all of the forces directly and then apply Newton's laws of motion. On this view, to ask "Which components of the system possess the additional energy in our example: $[q^4(\delta t)^2/2m][1/r^2R^2 - 1/R^4]$?" is like asking for the home address of the average family in the USA. The question is based on a mistaken conception of energy's reality.

The success of the conserved-quantities objection to retarded action at a distance thus depends ultimately upon energy's *ontological status*: the kind of thing energy is, the sense in which energy exists. ("Ontological" means "having to do with existence.") Perhaps a system's energy is a kind of stuff that is neither created nor destroyed and that cannot exist without being stored by or contained in something. The bits (or "parcels") of energy then have definite locations in space. Alternatively, perhaps energy (like velocity) is a property and so cannot exist except as the property of some thing (whose energy it is). In either case, the fact that electric and magnetic forces are retarded while energy is conserved gives us our objection to retarded action at a distance: there must be something to possess the additional energy needed to balance the books (in our example: $[q^4(\delta t)^2/2m][1/r^2R^2 - 1/R^4]$). This conception of energy's ontological status is implicit in Maxwell's famous statement of this argument in the final paragraph of his *Treatise on Electricity and Magnetism*:

[W]henever energy is transmitted from one body to another in time [there's the retardation!], there must be a medium or substance in which the energy exists after it leaves one body and before it reaches the other, for energy, as Torricelli remarked, "is a quintessence of so subtle a nature that it cannot be contained in any vessel except the inmost substance of material things." (1954: II, 493; cf. 1995: 365, 395, 812)

The most plausible repository of this energy is a field (though as we have just seen, there are other, ghostly possibilities).[3] Since some of the system's energy belongs to the field, the field must be real (Feynman et al. 1963: I, p.10-9).

On the other hand, energy may be merely a useful theoretical fiction, convenient shorthand for a certain algebraic combination of quantities. After all, that is what momentum often seems like. Admittedly, we might sometimes say that a moving body, in colliding with a stationary one, *transfers* some of its momentum to the second body by setting that body in motion. This "transfer" talk suggests that there is a certain *stuff* that continues to exist from one moment to the next; originally, all of it resided in one body, but later it was redistributed, some of it going into the second body. In other words, it sounds like what happens when there is a penny in my pocket, and then I give that penny to you: that selfsame penny is transferred into your pocket. But when we think of momentum as merely a certain algebraic combination of quantities – namely, mass times velocity – then we are less inclined to say that momentum is literally "transferred." Even in the special case where the velocity gained by the second body *equals* the velocity lost by the first, we would not say that the velocity was *transferred* – that *the very same velocity* was originally in the first body and then came into the possession of the second. The two velocities are equal, but a velocity differs from a penny in not being a *thing* with an identity, an individuality that it retains while moving from body to body. We can think of one body as losing momentum and of the other as gaining momentum without imagining that the same stuff once possessed by the first body is now possessed by the second, just as we can think of one person paying another – by credit card rather than cash – without imagining that there is some physical object that was transferred from the first person to the second. (The first person retains his credit card after the transaction.)

If energy is in this way merely a bookkeeping device, then there is no reason to agree with Maxwell that energy cannot be conceived without "a medium or substance in which the energy exists." An additional energy term that cannot plausibly be associated with any of the system's real (that is, non-ghostly) components could nevertheless be included in the formula for the system's total energy just because doing so is the only way to create a conserved quantity, and therefore is the proper way to "keep the books." Since this additional energy would not need to belong to some component of the system, there would be no argument here for the field's reality, and hence for spatiotemporal locality.

To investigate energy's *ontological status*, we must examine more carefully how energy is characterized in physics.

4 Energy in Classical Physics

In classical physics, there are two ways to think about the accumulation of force. One way is to think about the total force exerted *over an interval of time*, as when you push on a shopping cart from time t_1 to time t_2 as the cart rolls along a level surface. (For simplicity, let's assume that your push is the only force that the cart feels besides gravity and the equal and opposite upward force exerted by the floor.) Imagine dividing the interval from t_1 to t_2 into smaller pieces, each lasting Δt. For each of these small periods, imagine multiplying that period's duration by the average force F that you exerted on the cart in that period, and then adding all of these $(F\Delta t)$s for the entire interval from t_1 to t_2. As we make Δt smaller and smaller, the sum of these $(F\Delta t)$s converges to some value. This limiting value is the integral $\int F dt$ taken from t_1 to t_2. (The integral sign $\int$ is a stretched-out "S" for "sum.") According to Newton's laws of motion, this accumulated force equals the net change in the cart's momentum from t_1 to t_2. (For calculus fans, here's why: By Newton's second law ($F = ma = m\, dv/dt$), $\int F dt$ (taken from t_1 to t_2) $= \int m(dv/dt)dt = \int m\, dv = mv_2 - mv_1$, where v_2 is the body's velocity at t_2, and v_1 is its velocity at t_1.)

The second way to think about the accumulation of force is to think in an analogous way about the total force exerted *over a certain distance*, as when you push a shopping cart while it rolls from location r_1 to location r_2. Imagine dividing the path traversed by the cart from r_1 to

r_2 into smaller pieces, each of length Δr. For each of these pieces, imagine multiplying Δr by the average force exerted in the direction in which the cart rolls during that piece, and then adding all of these bits. (Each bit is the product of the average force vector $\boldsymbol{F}$ with the vector $\boldsymbol{\Delta r}$ having length Δr and pointing in the direction of that piece of the path.) As we make Δr smaller and smaller, the sum of these $(\boldsymbol{F} \cdot \boldsymbol{\Delta r})$s converges to a certain value. This value is the integral $\int \boldsymbol{F} \cdot \boldsymbol{dr}$ taken from r_1 to r_2. According to Newton's laws of motion, this result is the change in the cart's kinetic energy $(\frac{1}{2})mv^2$ from the start of the path to the end. (For calculus fans here's why: By Newton's second law ($\boldsymbol{F} = m\boldsymbol{a} = m\, \boldsymbol{dv}/dt$), the integral $\int \boldsymbol{F} \cdot \boldsymbol{dr}$ (taken from r_1 to r_2) $= \int m\, (\boldsymbol{dv}/dt) \cdot \boldsymbol{dr} = \int m\, \boldsymbol{dv} \cdot (\boldsymbol{dr}/dt) = \int m\boldsymbol{v} \cdot \boldsymbol{dv} = mv_2^2/2 - mv_1^2/2$.) Thus, momentum and energy are sibling concepts, both arising from expressions for the accumulation of force – over time and space, respectively.[4]

The above calculation, showing $\int \boldsymbol{F} \cdot \boldsymbol{dr}$ (taken from r_1 to r_2) to equal the change in the cart's kinetic energy, presupposed the force *you* exerted on the cart to be the *total* force on the cart in the direction it was moving. (That is because the calculation took the force you exerted to be related, by $\boldsymbol{F} = m\boldsymbol{a}$, to the cart's *net* acceleration.) Suppose that rather than pushing the cart across a level surface, you were pushing the cart *uphill*. Suppose the cart began and ended its path at rest. Then the cart's kinetic energy would be unchanged even though force had worked upon it through a distance: taken from r_1 to r_2, $\int \boldsymbol{F} \cdot \boldsymbol{dr} \neq 0$. This is why the notion of "potential energy" (energy that reflects position rather than speed) must be introduced: the force accumulated over a distance – or "work," as physics textbooks say – needed to push the shopping cart uphill from r_1 (at rest) to r_2 (at rest) is the change in the cart's *total* energy, kinetic plus potential. The general principle is that a system's total energy is the work (force accumulated over a distance) needed to assemble the system: to put its various parts into their positions and to give them their velocities.

Potential energy, then, is just as *real* as kinetic energy. (Potential energy is not energy that exists merely *potentially* rather than *actually*.) The significance of a system's total energy is that it remains constant as long as the system is closed (that is, isolated). When the shopping cart, at rest at r_2, is allowed to roll back down the slope to r_1, it picks up speed. Its total energy remains constant (disregarding friction with the ground, air resistance, and so on – that is, approximating the cart as a closed system). The potential energy it loses in the descent equals the kinetic energy it gains.[5]

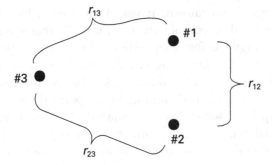

Figure 5.2   Three charged bodies at rest.

A system's energy is the work needed to assemble it. Imagine the system as three point bodies at rest, each with some positive charge (q_1, q_2, and q_3) and separated from the others by various distances (figure 5.2). Let us begin the system's assembly by bringing q_2 from infinitely far away to within r_{12} of q_1. We apply a force to q_2 that is just enough to balance the electrostatic repulsion between q_1 and q_2. This will make q_2 move toward q_1 infinitesimally slowly. So when it arrives at its proper place, q_2 will be at rest there and Coulomb's law gives the repulsion between the two charges when they are separated by a distance r: $q_1 q_2 / r^2$. So the work needed to bring in q_2 is the integral $\int \boldsymbol{F} \cdot \boldsymbol{dr}$ (taken from infinity to r_{12}) $= \int -[q_1 q_2 / r^2] dr = q_1 q_2 / r_{12}$.[6] Next q_3 is brought into position from infinitely far away. It is repelled by both q_1 and q_2, so we must do two integrals, yielding $q_1 q_3 / r_{13}$ and $q_2 q_3 / r_{23}$. So the system's total potential energy is $q_1 q_2 / r_{12} + q_1 q_3 / r_{13} + q_2 q_3 / r_{23}$.

There are two features of this result that are relevant to energy's ontological status. First, this calculation assumed that the system is assembled from an initial state in which each point charge lies infinitely far away from each of the others; work is done in moving the various charges together from infinity. This work equals the *difference* between the energy of the system *assembled* and its energy *initially* (as "raw materials" infinitely remote from one another). This energy *difference* equals the assembled system's *total* energy only if the energy of the system as raw materials was zero. But no justification was offered for treating a stationary charge at infinity as having zero energy. In other words, all that the above calculation determined was the energy of the assembled system *relative to* the energy of its components when infinitely separated – that is, relative to the energy of what we considered to be the raw materials out of which the system was built. The *absolute* value of the system's energy is not known. It would have to include the energy needed to put together each of those raw materials (that is, each charged body's "self-energy") before they were used to assemble the system.

We do not need to know the system's absolute energy in order to use the principle of energy conservation. To see why, suppose the charged bodies when at rest and infinitely separated (the "raw materials") actually have some non-zero quantity C of energy. Then the system's actual total energy at any moment differs by C from the energy we calculate by mistakenly presuming the system to have zero energy when the charges are at rest and infinitely separated. But no harm done: since the total energy is conserved, and the total energy is always C plus our calculated energy, it follows that our calculated energy must be conserved.

That we can know only energy *differences* – that energy's absolute value would be a "dangler," making no difference to anything else – suggests (though falls well short of showing conclusively) that there is *no fact* of the matter regarding the absolute value of a system's energy. (That it does not exist would nicely explain why it makes no difference.) On this view, energy is not real stuff and absolute energy content is not a real property of things. We have already examined the strengths and weaknesses of this sort of argument, since in chapter 2, we encountered an argument of the very same kind applied to the electric *potential*: just as it is arbitrary what to count as the "raw materials" out of which the system is assembled, so likewise the zero of potential (that is, the "sea level" when potential is thought of as elevation) is arbitrary. This similarity is no coincidence: the electric potential at a given location is the potential energy that a body there would have, for each unit of its charge – that is, the work needed to move a unit electric charge to that location from the location of zero potential (the "ground"), which must also be the location at which a body has zero potential energy. "Potential energy" denotes energy associated with a potential (Maxwell 1892: 143); if 5 ergs is the potential energy of a unit charge at a given location, then 5 ergs per statcoulomb is the potential there. Both potential and potential energy are specified relative to a location that we designate for convenience as where they are zero. This suggests that the absolute values of these quantities correspond to nothing in reality – in other words, that we can make no factual error by taking the zero to be at one location rather than another.

On this view, energy has the same ontological status as certain other physical quantities. Take valence in chemistry. On a simple valence theory, a combination of atoms forms a stable chemical compound if and only if the valence numbers of those atoms sum to zero. For instance, oxygen's valence is −2 and hydrogen's is +1, so water (H_2O) is stable.[7] Now valence numbers would predict exactly the same stable chemical compounds even if every valence number were multiplied by (say) 3. For example, the valence numbers of two hydrogen atoms (3 + 3) and one oxygen atom (−6) would still add to zero. Only the *relative* values of the valence numbers (in particular, their ratios) make a difference, not their *absolute* values. This makes it more difficult to think of hydrogen's valence number as truly being +1 rather than +3 – that is, to interpret its valence number *literally* as the *number of something*, such as the number of hooks that a hydrogen atom possesses. (Likewise, it becomes more difficult to think of oxygen's valence number as −1 times the number of eyes that an oxygen atom possesses, where a hook latches on to an eye.) An alternative interpretation of an atom's valence number would be as the ratio of its number of hooks (or, for negative valence numbers, its number of eyes times −1) to the number of hooks possessed by a hydrogen atom. Or perhaps valence is just a useful scheme for predicting, remembering, and systematizing which compounds are stable. Our question is whether energy is also merely a theoretical device.

The above calculation of the three electric charges' potential energy has a second feature relevant to energy's ontological status. We began our calculation by bringing q_2 from infinity to within r_{12} of q_1; we then moved q_3 into place. The work done in moving q_2 was q_1q_2/r_{12} and in moving q_3 was $q_1q_3/r_{13} + q_2q_3/r_{23}$. This suggests that in our final arrangement, the potential energy of q_1 (which is its total energy, since it is stationary) is zero, that of q_2 is q_1q_2/r_{12}, and that of q_3 is $q_1q_3/r_{13} + q_2q_3/r_{23}$. However, imagine that instead we had first brought q_3 from infinity to within r_{23} of q_2, and then moved q_1 into place. The *total work* would have been the same, although the work needed to bring in each individual charge would plainly have been different: the potential energy of q_2 would have been zero, that of q_3 would have been q_2q_3/r_{23}, and of q_1 would have been $q_1q_2/r_{12} + q_1q_3/r_{13}$. What this means is that the current arrangement of the system's components determines the system's current total potential energy (setting aside the energy of the raw materials) but fails to determine the potential energies of its various parts.

In fact, the use we make of the concept of energy (in connection with energy conservation) does not require that we know how the system's energy is distributed among its components. That is because those components are interacting with one another and so do not individually constitute closed systems; energy conservation applies only to a closed system. Since energy's distribution among a system's components is not determined by the system's current state and makes no difference to any fact regarding the system's state at any other moment, it is a dangler. This suggests to some degree (by the same kind of argument we have seen before) that there is no fact of the matter regarding the energy possessed by any of the system's parts. There is only the energy of the system as a whole. This view is often found (at least implicitly) in physics textbooks. When they discuss the shopping cart at the top of the slope, they refer not to the *cart*'s potential energy, but to the potential energy of the *system* consisting of the cart and the Earth. On this view, energy is not a stuff (since it has no particular distribution in space), and if energy is a property, then it differs from mass, charge, and momentum in that a system's energy is *not* the sum of the energies of its parts.[8]

In short, although the concept of energy "works," there are obstacles to understanding energy as a real stuff or property. Although physics textbooks today do not emphasize these obstacles, physicists and philosophers have long grappled with them. For if energy is merely a bookkeeping device, then it cannot be used to show that fields are real, and so that spatiotemporal locality holds. To combat the tendency of physics textbooks to desensitize us to the difficulties that energy presents, I offer the following remarks from Sir Oliver Lodge (English physicist, 1851–1940):

> [T]he idea of "potential energy" has always been felt to be a difficulty. It was easy enough to take account of it in the formulae, but it was not easy or possible always to form a clear and consistent mental image of what was physically meant by it.
>
> A stone is raised, it gains potential energy; but how does the stone "up" differ from the stone "down"?, and how can an inert and quiet stone be said to possess energy? Well, then, the stone hasn't the energy but the earth has, or rather "the system of earth and stone possesses energy in virtue of its configuration." True, but foggy.   (1885: 484)

The stone, after having been raised to a great height, seems intrinsically no different from the stone as it was on the ground, yet its energy is supposed to be different. This is not so puzzling if the potential energy is not an intrinsic property of the stone, but is a property of the stone–Earth system. However, it is then difficult to see energy as stuff having a definite location. We shall see shortly that these "philosophical" worries were fruitful stimulants to the development of field theory in the nineteenth century. These developments seem to dispel some of the "fog" Lodge notes, but also raise new puzzles.

5 Energy in the Fields

At the start of this chapter, we saw that retarded action at a distance poses a problem for energy conservation, and that one way to avoid this problem is to regard the field as possessing energy and therefore real. On this interpretation, as we shall now see, the field's energy is the energy we regarded in the previous section as the electric potential energy. This reinterpretation of potential energy helps to resolve some of our worries about interpreting energy as real: the field's energy turns out to be assigned a definite distribution in space and to be calculated relative to a non-arbitrary zero level. However, this interpretation also ultimately raises questions that might make us hesitate to regard energy as real and possessed by the field.

To see how energy can be assigned to the field, let's look at a standard example: a hollow sphere of radius R at rest possessing a total charge q uniformly distributed on its surface. When you sum all of the contributions that the sphere's charges make to the electric field at a given point, you arrive at a famous result. At any point inside the sphere, the field is zero (figure 5.3). At any point outside the sphere at a distance r from its center, the field is q/r^2 pointing away from the sphere's center – exactly the same as the field of a point of charge q located at the sphere's center (figure 5.4). Let us use this result to calculate the sphere's electric potential energy. (You can ignore the mathematical details of this calculation, if you wish. The general idea is all that you need to grasp.)

Its electric potential energy equals the work needed to assemble it by bringing its parts in from infinity at infinitesimal speed. Imagine this process as like taking a very large sphere and pressing its parts together so that it becomes a smaller and smaller sphere, eventually contracting to radius R. Let us look at one stage of this process, when the sphere had not yet been reduced all the way to radius R; at this moment, it has some larger radius r. Just inside the spherical surface,

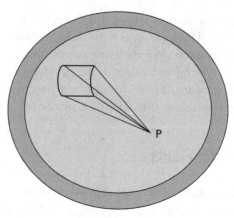

Figure 5.3 A hollow, uniformly charged sphere at rest, shown in cross section. Consider the electric field at interior point P. Divide the sphere's surface into small pieces of equal area. What contribution to that field is made by one of these pieces (such as the one shown, with lines connecting its corners to P)? By $E = q/r^2$ (equation 2.6), that piece's contribution is proportional to the piece's total charge q, which (since the sphere's surface is uniformly charged) is proportional to the piece's surface area, which is proportional to r^2, where r is the piece's distance from P. But $E = q/r^2$ also says that the piece's contribution to E at P is inversely proportional to r^2. These two proportionalities cancel, so every piece's contribution to $\boldsymbol{E}$ at P is the same magnitude. But since the contributions of the sphere's various pieces point in all directions, they cancel each other, leaving $\boldsymbol{E} = 0$ at P.

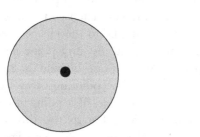

Figure 5.4 The sphere's field at P is the same as the field at P of an imaginary point body bearing charge q and located where the sphere's center is.

$E = 0$. Just outside, $E = q/r^2$. So at the surface itself, E is the average of these: $q/2r^2$. The electric force repelling a piece of the surface from the rest of the sphere (figure 5.5) equals the piece's charge times the field there, according to $\boldsymbol{F} = q\boldsymbol{E}$ (equation 2.5). A piece of infinitesimal surface area dA has charge $(q/4\pi r^2)dA$. (This is the piece's area times the density of charge on the sphere's surface, which is the sphere's

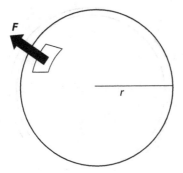

Figure 5.5 The black arrow shows the force ***F*** on a piece of the surface, repelling it from the rest of the sphere – the force that must be opposed in order to compress the sphere further.

total charge q divided by its total surface area $4\pi r^2$.) So by ***F*** $= q$***E***, this piece feels a force (resisting its further compression) of strength $[(q/4\pi r^2)dA]q/2r^2 = q^2 dA/8\pi r^4$. To compress the sphere (at infinitesimal speed), we must balance this outward force by applying an equal inward force. So the work done in moving a part of the sphere inward by a small distance dr is $-[q^2 dA/8\pi r^4]dr$.[9] Each piece of the sphere requires the same force, so since the sphere's total surface area is $4\pi r^2$, the total work done in diminishing the sphere's radius by dr is $-[q^2/8\pi r^4]4\pi r^2 dr$, which simplifies to $-[q^2/2r^2]dr$. That's just one stage of the process of assembling the sphere. The entire process diminishes its radius from infinity to R. Adding the work done in each stage, the entire process does work totaling $\int -[q^2/2r^2]dr$ (taken from infinity to R) $= \int [q^2/2r^2]dr$ (taken from R to infinity) $= q^2/2R$. A pretty result for the sphere's electric potential energy!

Now watch this: At any point in space, E has some value, and so we may consider the value of $E^2/8\pi$ there. (Why have I chosen this particular quantity? You'll see in a moment.) This quantity equals zero at any location within R (where ***E*** $= 0$), and it equals $(q/r^2)^2/8\pi = q^2/8\pi r^4$ at any location beyond R. Let us think of $E^2/8\pi$ as the quantity per unit volume (the density, in other words) of some stuff sitting in space. Let's figure out the total amount of this stuff in all of space. To begin with, we can figure out how much of this stuff is contained in the region between two imaginary spheres, with radii r and $r + dr$, concentric with our actual sphere (figure 5.6). The volume of this region is $4\pi r^2 dr$. If $r < R$, the region holds none of the stuff since $E = 0$ there. If $r > R$, the stuff's density is the same throughout the volume,

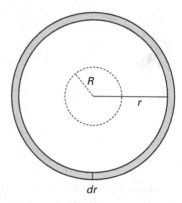

Figure 5.6 A region of width *dr* between two imaginary spheres, with radii *r* and *r + dr*, concentric with our actual hollow sphere of radius $R < r$.

equaling $q^2/8\pi r^4$ as we just saw. So the quantity of stuff (density times volume) in the region between the two imaginary spheres is $[q^2/8\pi r^4]4\pi r^2 dr = q^2 dr/2r^2$. To calculate how much stuff there is in all of space, we must sum the contributions made by each of these regions from R to infinity: $\int [q^2/2r^2]dr$. Look familiar? Sure: this is exactly our result in the previous paragraph for the system's electric potential energy. The "stuff" can be thought of as energy! (It can be proved that these two calculational procedures will give the same answer in any static case – Rosser 1997: 282–4; Knudsen 1985: 160–4.) We can interpret the work done in assembling a configuration of stationary charges as having been spent in building up their electric field rather than in moving those charges in from infinity.

Here is the upshot: In the previous section, we conceived of the energy possessed by an arrangement of charges as consisting of the charges' kinetic energies plus the potential energy of the arrangement as a whole. Now we have seen that we can equally well think of the potential energy as the energy belonging to the arrangement's electric field. The field's "energy density" (its energy per unit volume) is $E^2/8\pi$. That is, if we have a volume V in which the field is uniformly E, then the field in that volume possesses a total energy of $E^2/8\pi$ times V. The field's total energy in a region of space is the sum of its energies in each part of that region. That sum is $\int [E^2/8\pi]dV$ taken over the region.[10]

This interpretation assigns a definite distribution in space to the potential energy. Rather than belonging *somehow* to the arrangement of charges *as a whole*, the energy is spread around with density $E^2/8\pi$. Now consider the other obstacle I mentioned earlier to interpreting energy as real stuff: the arbitrariness of energy's zero level. At first glance, it might appear that the same arbitrariness arises if we think of the potential energy as distributed in the field. Suppose we took the field's energy density to be $E^2/8\pi$ *plus* some arbitrary quantity C, where C's contribution to the field's total energy (in other words, $\int C \, dV$ taken over all space) would remain the same (say, 5 ergs) no matter what. (So C cannot be a function of E, since then $\int C \, dV$ taken over all space would surely change if, say, $E = 0$ everywhere.) Let's see why this arbitrary constant in the energy density would make no difference to energy conservation. Suppose that the total energy, calculated using $E^2/8\pi$ as the field's energy density, is conserved. This is always 5 ergs less than the system's total energy calculated using $E^2/8\pi + C$ as the energy density. Therefore, the system's total energy, calculated using $E^2/8\pi + C$, must be conserved. Apparently, then, we could just as well take the field's energy density to be $E^2/8\pi + C$, and so the arbitrariness of energy's absolute level remains.

However, it seems very sensible to suppose that wherever $E = 0$, the field's energy density is zero: there can be no energy stored by the field at a location where there is no field! So wherever $E = 0$, it must be that $C = 0$. But then wherever $E \neq 0$, C must still be 0, since C is not a function of E. Hence, $C = 0$ everywhere under any circumstances. Energy's zero is not arbitrary because the field's zero is not arbitrary!

In other words: The potential energy is the energy required to assemble the system out of its raw materials, but the energy required to make those raw materials is not included. If a system's potential energy belongs to its bodies ($q_1 q_3 / r_{13} + q_2 q_3 / r_{23} + \ldots$), then the potential energy has an *arbitrary* zero because we can imagine *any* condition to be the starting point from which the system is assembled. On the other hand, if the system's potential energy is interpreted as its field's energy, then the system's potential energy has a *natural* zero: namely, when the *field* is zero. In constructing the system's field from an initial $E = 0$ condition everywhere, we are assembling the system from a non-arbitrary starting point.

Having thus removed the earlier obstacles to thinking of energy as real stuff, we seem to be able to think of the field as real in virtue of possessing energy. As Lodge says (continuing the passage on the hazy concept of potential energy):

> When universal contact action [that is, spatiotemporal locality] is admitted, the haze disappears; the energy is seen to be possessed, not

> by stone or by earth or by both of them, but by the medium which
> surrounds both and presses them together . . .   (1885: 484)

And if the field is real (in virtue of possessing energy), then spatio-temporal locality is upheld.

Of course, Lodge's remark began by *presuming* spatiotemporal locality, so the line of argument he traces cannot be used to *demonstrate* spatiotemporal locality. We cannot use spatiotemporal locality to argue that energy is real and belongs to the field, if we are then going to turn around and use these facts to show that spatiotemporal locality holds. We still lack an *independent* reason to believe energy real rather than merely a calculational device. We can rebut objections to energy's reality by ascribing energy to the field rather than the charged bodies. But that is no argument for believing energy real, since we did not already have good independent reasons for believing the field real.

In *static* cases (such as the hollow sphere at rest), the integral of $E^2/8\pi$ over all space (assigning the energy to the field, with a definite distribution) gives the same total electric energy as does the sum of $q_1 q_3 / r_{13} + q_2 q_3 / r_{23} + \ldots$ (assigning the energy to the system of charged bodies as a whole). But the latter method fails to secure energy conservation in *dynamic* cases, as we learned in section 2. On the other hand, the energy-density formula from Maxwell's electromagnetic theory ($E^2/8\pi$) assigns the fields just the right amount of energy to balance any change in the bodies' total kinetic energy, allowing energy conservation to hold. So in computing a system's energy, even a friend of retarded action at a distance would have to use $E^2/8\pi$ in order to make the books balance. Of course, she could not interpret this formula as representing the field's energy. As we saw earlier in this chapter, the easiest way for her to co-opt this formula would be to interpret the potential energy not as real and belonging to the bodies, but as merely a calculational device. On that view, the integral taken over all space does not give the total quantity of some real stuff distributed in space. It is just the right calculation for predicting the system's behavior by using the law of energy conservation.

6   Energy Flow and the Poynting Vector

Suppose two oppositely charged bodies, initially held at rest, are released. They accelerate toward each other in virtue of their mutual electric attraction. Each body gains kinetic energy. Before we ascribed energy to the field, we might have said that some of the configuration's electric potential energy is transformed into kinetic energy. This raises several difficult questions if energy is something real: not only what it means for energy to undergo this "transformation," but also where the potential energy was before it came to be possessed by the bodies as kinetic energy. If it had a definite location, where was it and how did it manage to get from there to the charged bodies? If it did not have a definite location but instead was possessed by the pair of bodies *as a whole*, then how did it manage to acquire a definite location (in coming into the possession of a body as its kinetic energy)?

These questions take a different form if we interpret the electric potential energy as belonging to the electric field. As the two bodies accelerate, the field's energy diminishes and each body's energy increases. Can we come up with a consistent picture of this process? (If not, then it appears that we must think of "energy" as merely a calculational device.)

Suppose that energy is some kind of stuff. When the field's energy diminishes and the bodies' energy increases, does some of this stuff, existing somewhere in the field, suddenly cease to exist, and an equal amount of this stuff simultaneously come into existence in the possession of one of the bodies? This would be consistent with energy conservation, since the total amount of energy would remain the same. It would also be consistent with temporal locality: there would be no gap in time between the disappearance of one bit of energy and the appearance of another. But this picture would apparently be inconsistent with spatial locality, since the body's location (where the energy comes into existence) might be far away from where the energy disappears in the field. To uphold spatiotemporal locality, we must think of the energy-stuff as *moving continuously*: flowing from distant regions of the field into regions closer to the body, and thence into regions adjoining the body, and ultimately into the body.

Heaviside made this point elegantly:

> The principle of the continuity of energy is a special form of that of its conservation. In the ordinary understanding of the conservation principle it is the integral [in other words, the total] amount of energy that is conserved, and nothing is said about its distribution or its motion. This involves continuity of existence in time, but not necessarily in space also. But if we can localize energy definitely in space [that is, ascribe it

to the field, with a distribution given by $E^2/8\pi$], then we are bound to ask how energy gets from place to place. If it possessed continuity in time only, it might go out of existence at one place and come into existence simultaneously at another. This is sufficient for its conservation. This view, however, does not recommend itself. The alternative is to assert continuity of existence in space also, and to enunciate the principle thus: When energy goes from place to place, it traverses the intermediate space. (1922: I, 73–4; cf. Feynman et al. 1963: II, pp.8-10 to 8-11, and 27-1 to 27-2)

Hey, that's a catchy slogan! This view presupposes that some bit of energy at one place and time may be *the selfsame bit* as some bit at another place and time; parcels of energy have continuing *identities* through time. Otherwise, energy would not really be *moving* or *flowing* or *transferred* from the field to a body, since these notions require that *the same entity* that is here now have been there then. Lodge entitled the paper from which I have twice quoted "On the Identity of Energy":

> This notion is, I say, an extension of the principle of the conservation of energy. The conservation of energy was satisfied by the *total quantity* remaining unaltered; there was no individuality about it: one form might die out, provided another form simultaneously appeared elsewhere in equal quantity. On the new plan we may label a bit of energy and trace its motion and change of form, just as we may ticket a piece of matter so as to identify it in other places under other conditions; and the route of the energy may be discussed with the same certainty that its existence was continuous as would be felt in discussing the route of some lost luggage which has turned up at a distant station in however battered and transformed a condition. (1885: 482)

Those quaint Victorian analogies! (Actually, this remark is not at all outdated, despite the shift from rail to air travel.)

Physicists often seem to trace an energy parcel's career, thereby presuming it to have a continuing identity. They might say that some energy supplied by the battery was conveyed to the wire, made the current go, and was later dissipated as heat (as the wire warmed up in resisting the current's flow), being transferred ultimately to the surrounding air molecules as kinetic energy. We must see whether electromagnetic theory can be interpreted as giving us a precise, intuitive picture of the rate and direction of energy's flow at every location in

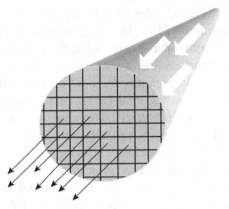

Figure 5.7   Energy flowing down the pipe in accordance with the white arrows. There are patches over the surface covering the pipe's exit, and a few of those patches have their vectors **S** shown (black arrows).

the field. At a minimum, we must see whether electromagnetic theory allows every point in space to be assigned a vector **S** pointing in the direction in which the energy there is flowing and having a magnitude equal to the rate at which energy there is flowing in that direction.

To see what this means, imagine a pipe down which energy is flowing like water through a hose (figure 5.7). Imagine stretching a surface tightly over the pipe's exit (like covering the end with plastic wrap). Divide that surface into small patches. Associated with each patch, there is a vector **S**. The component of **S** perpendicular to the patch equals the amount of energy that flows out of the pipe and thence over every square centimeter of the patch each second. In other words, the amount of energy that flows over the patch in 1 second is the patch's surface area (in square centimeters) multiplied by **S**'s component perpendicular to the patch. If we allow the patches to become infinitesimally small (so that each has surface area dA), then we can associate a vector **S** with each *point* on the surface stretched over the pipe's exit. The rate at which energy flows out of the pipe equals the sum, for each of these points, of **S**'s component perpendicular to the surface at that point multiplied by the infinitesimally small surface area dA: the integral $\int \boldsymbol{S} \cdot d\boldsymbol{A}$ over the surface covering the end of the pipe. This is called the "flux" (Latin for "flow") of energy exiting the pipe, and **S** at a given point is the "energy flux density" there: the rate at which energy flows across that point. It comes in units of energy per square centimeter per second.

What must S be like in order for it to depict energy as some kind of stuff flowing continuously? Imagine a region of space (empty of matter) containing electric and/or magnetic fields; the boundary of the region is a certain closed surface (like a soap bubble floating in space). The total field energy in the region is the total energy in the electric field there ($\int [E^2/8\pi]dV$) plus the total energy in the magnetic field there (which, analogously, is $\int [B^2/8\pi]dV$). If energy flows continuously, then the total energy in the region can change only by virtue of energy flowing in or out over the surface enclosing the region. So the total flux of energy passing out of the region through the surface enclosing it ($\int S \cdot dA$ – a negative quantity if the net flow is *into* the region) equals the rate at which the region's energy ($\int [1/8\pi][E^2 + B^2]dV$) is decreasing. This is the *continuity equation* for energy. Is there an S satisfying the continuity equation? Yes: In 1884, John Henry Poynting (English physicist, 1852–1914) discovered that this equation is solved by $S = (c/4\pi)E \times B$. This S is called the "Poynting vector." (Isn't it delicious that the *Poynting* vector *points* in the direction of energy flow?) The "$\times$" stands for "cross product" vector multiplication, which I introduced in chapter 2 (figure 2.6).

If we are going to support spatiotemporal locality by regarding energy as real stuff that moves continuously, then apparently, we must take the Poynting vector seriously. To see whether the Poynting vector can plausibly be interpreted as energy flux density, let's look at its direction in some examples. Consider a van de Graaf generator (figure 5.8), by which a great electric charge can be accumulated on a metallic sphere (labeled D in the figure). Someone turns a pulley, thereby moving a vertical conveyer belt, which rubs against fine needle points (at A), which attract and remove electrons (negatively charged) from the belt. The belt carries the resulting positive charges up to the sphere, where more needle points (at B) allow some of the sphere's electrons (attracted to the positively charged belt) to pass to the belt from the sphere, increasing the sphere's positive charge. Later, positively charged particles from the sphere accelerate through an evacuated tube (C), arriving at the ground with high kinetic energy. Intuitively, how does energy flow through this system? Mechanical energy enters through the pulley moving the belt that carries positive charges upward toward the positively charged sphere, adding to the electric potential energy. (Since like charges repel, it takes work to move more positive charges toward the sphere.) We have interpreted this electric potential energy

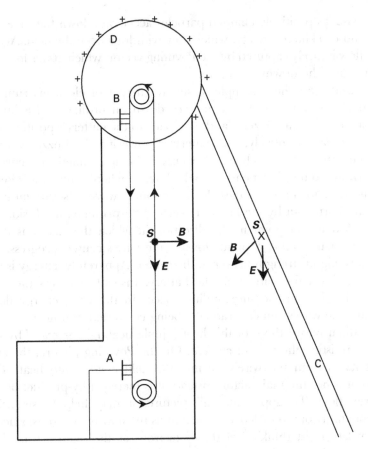

Figure 5.8   A van de Graaf generator. The **E** field lines point away from the concentration of positive charge on the globe at the top. The **B** field lines loop around the charge-carrying conveyer belt. At a location on the belt's facing surface, vectors **E** and **B** are shown, which make vector **S** (shown as a dot) point out of the belt towards you, dear reader. Also, at a location on the facing surface of the discharge tube (slanting off diagonally downward and to the right) containing a near vacuum, **E** and **B** are shown, which make **S** (shown as an X) point into the tube.

as field energy. So as positive charges travel on the belt from A to B, they add energy to the surrounding field. Indeed, on the belt's surface facing us, the Poynting vector $S = (c/4\pi)E \times B$ points outward from the belt – toward us, into the surrounding space where the field lies. (On the rear surface of the belt, the Poynting vector is again pointing away from the belt – away from us, into the surrounding space.)

Likewise, as positively charged particles accelerate down the tube (C), they absorb kinetic energy, which they withdraw from the field. Again, this flow is nicely captured by the Poynting vector, which at this location points into the downward current.

Consider another example: a steady current of electrons running from the negative pole of a battery down an ordinary wire into a resistance – say, a buzzer – and then back to the battery's positive pole (figure 5.9a). Intuitively, the battery is powering the buzzer via the current: the battery's chemical energy is being drained, the current is carrying some of that energy to the buzzer, where some of it is being transferred to air molecules, forming sound waves. Is this intuitive picture borne out by the Poynting vector? Surprisingly, no: Inside the wire, S does not point in the direction in which the electrons flow. Rather (figure 5.9b), S points toward the wire's center, progressively diminishing in strength nearer to the center. Apparently, energy is not flowing down the wire from the battery. Instead, it enters the wire from the field in the surrounding space. In the wire, energy flows toward the wire's center, gradually being converted into heat.

Ordinarily, we think of this heat ("Joule heating") as *caused* by the wire's resisting the flow of current. On the Poynting picture, the current *results* from the wire's turning the field energy into heat. The current is not the main actor; instead, it is merely a by-product of the energy flow. To appreciate this picture, it may help to set aside momentarily our knowledge of current as the flow of electrons, since it is hard to resist thinking of these electrons as the main actors. The nature of electric current was unknown to Maxwell and his followers, such as Poynting. To them, energy was the more understandable,

---

Figure 5.9 (*opposite*)

(a)   A circuit running from a battery's negative pole through a buzzer and back to the battery's positive pole. Arrows depict the path of the electrons down the wire.

(b)   Inside the wire, shown in cross section. E is uniform and points down the wire toward the battery's negative pole (since that is the direction in which a positive test charge would be attracted). B is circular around the current and increases with distance from the wire's center. Do $E \times B$ yourself to verify that S everywhere in the wire is pointing deeper into the wire. After Skilling (1948: 133).

(c)   Some of the circuit's magnetic lines of force are shown.

(d)   The electric field near the wires is shown. The extent to which the field departs from being at right angles to the wire has been exaggerated.

(e)   The Poynting vector is shown. The extent to which it departs from being parallel to the wires has been exaggerated.

(a)

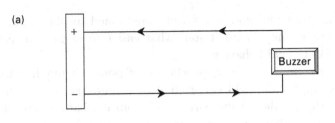

(b)

Path of electrons
down the wire

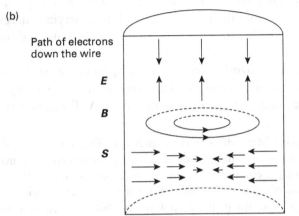

(c)

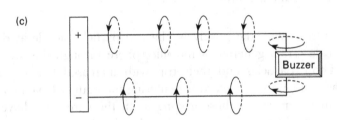

(d)

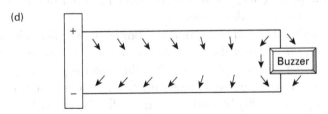

(e)

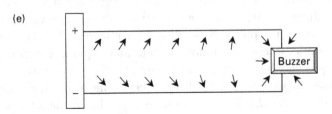

unifying principle, and fields were envisioned as the basic actors. Accordingly, they found it somewhat easier to regard current as a side-effect of field behavior.

Outside the wire (see figure 5.9c,d,e), *S* points slightly into the wire (accounting for the field energy that turns into heat within the wire) and mostly parallel to the wire, away from the battery, and into the buzzer. On this picture, the buzzer is drawing its energy not from the current, but from the field. The wire is not carrying energy to the buzzer, but is guiding the energy flowing from the battery through the field to the buzzer.

If we take the Poynting vector seriously, then, we must not regard the electrons in the current as carrying the battery's energy to the buzzer. The energy travels outside the wires. As Poynting wrote:

> [I]f we accept Maxwell's theory of energy residing in the medium [the field], we must no longer consider a current as something conveying energy along a conductor. A current in a conductor is rather to be regarded as consisting essentially of a convergence of electric or magnetic energy from the medium upon the conductor and its transformation there into other forms.   (1920: 192)

This contrasts with the positively charged particles accelerated from the van de Graaf generator's metallic sphere through the evacuated tube. Those particles end their trip with increased kinetic energy, whereas in a circuit, the electrons entering the battery have about the same kinetic energy as those leaving it. In the wire, they have converted the field's energy into heat, but they have not bestowed any of their own energy on the buzzer.

Admittedly, it is surprising that the buzzer is supplied with energy from everywhere *except* the wire. Nevertheless, we may be willing to bite the bullet and accept this result rather than reject the picture of energy as real stuff flowing through the field, a picture that supported spatiotemporal locality by underwriting the field's reality. But consider another example: a stationary electric charge near a stationary bar magnet (figure 5.10). At various locations, the charge contributes *E* and the magnet supplies *B*, so there is a non-zero *S*. The system is entirely static, so energy cannot be accumulating in any location. The Poynting vector bears out this intuition: over any *closed* surface (which you could picture as a soap bubble enclosing a volume), the total energy flow ($\int S \cdot dA$) is zero, and so in any volume, the total energy is

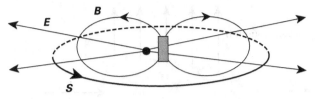

Figure 5.10   The dot is a stationary positive electric charge (with **E** lines emanating from it) near a bar magnet (with **B** lines emanating from its north pole and re-entering the magnet at its south pole). The Poynting vector **S** depicts energy flowing round and round. **S** comes out of the page on the left side and enters the page on the right side.

unchanging. However, **S** is non-zero in various locations, so *energy is flowing*. Take the quantity of energy that flows *into* a given region through *some* of its sides. (Those sides form an "open surface," which differs from a "closed surface" – like a soap bubble – in that an open surface *fails* to *completely* enclose a volume.) According to the Poynting vector, that inward-flowing quantity of energy is perfectly balanced by the quantity flowing *out* of the volume through its *other* sides (figure 5.11). So this perpetual flow of energy, round and round (figure 5.10), makes no difference to the quantity of energy in any region. That quantity is steady, and all of the bodies are unmoving – so why should we believe that any energy is flowing at all? Should we take the Poynting vector seriously here?

I recall asking myself this question in an undergraduate physics class. My textbook said:

> It is convenient to consider Poynting's vector as representing a flow of energy. Although this interpretation is usually correct, it is not rigorous, as can be seen by superposing arbitrary [electro]static and magnetic fields [as in our example]. In such a field, [**S** is non-zero in some places] despite the fact that there is, of course, zero energy flow everywhere. (Corson and Lorrain 1962: 321)

This remark can only leave a student asking: How am I supposed to know when I should take the Poynting vector seriously ("this interpretation is *usually* correct") and when I should not? Should the Poynting vector be interpreted as the energy flux density in the case involving a current through a buzzer, where (as we just saw) this interpretation leads to somewhat strange results? Alas, the remark from my old textbook is not unusual. A more recent text says:

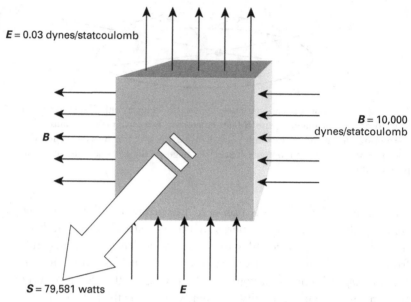

E = 0.03 dynes/statcoulomb

B = 10,000
dynes/statcoulomb

B

S = 79,581 watts   **E**

Figure 5.11   Here is a cubical region of space (one cubic cm in volume) in which there are uniform perpendicular **E** and **B** fields. The **E** field is modest in strength, and the **B** field could easily be reached between the poles of an iron magnet. According to the Poynting vector, there should be a considerable quantity of energy (its flow measured in watts, that is, joules of energy per second) flowing in through the region's back surface and out towards us through its facing surface. (By comparison, an ordinary household light bulb emits energy at a rate of merely 60 watts.) The six sides of the cube all together form a "closed surface." Any single side or combination of sides less than all six form an "open surface." Since the flow into the cube through the rear surface equals the flow out of the cube through the front surface, there is no net change in the amount of energy contained in the cube. (After Jeffries 1992: 387.)

Strictly speaking, only the integral of [**S**] over a closed surface has been shown to represent the energy flow, but in most cases, [**S**] does represent the flow of energy per unit area at each point. An obvious exception to this would be a region with an electrostatic and a steady magnetic field arising from different sources . . .   (Bleaney and Bleaney 1976: 233–4)

Here's another:

[A]lthough the integral of [**S**] over a surface may commonly be taken to represent power flow, this idea may sometimes lead to absurd inter-pretations (i.e., if the electrostatic field and the field of a permanent magnet exist in the same region). But the integral of [S] over a *closed* surface is always the true outward flow of energy.   (Skilling 1948: 133; see also Larmor 1897: 285)

But this "mixed" interpretation of S might well seem unstable. If we find the endless circulation of energy in static cases "absurd," then perhaps we should take these examples as demonstrating that for *any* surface, closed or open, we would be mistaken in interpreting S as the flow of energy stuff over that surface. Rather than interpreting S's flux over a closed surface and over *certain but not all* open surfaces to be energy flow, shouldn't we simply deny that the Poynting vector *ever* captures the flow of some real stuff – by denying that energy is such a thing?

On this basis, we might take energy to be merely a useful calculational device. Since the alleged circulation of energy fails to alter the energy density anywhere in the field, it cannot be detected by a device measuring changes in the energy distribution. Rather, a device would have to absorb some of the energy as it flows through, thereby converting that field energy to energy of some other form. However, there would then be less energy in the fields, so E and B would have to change, despite the fact that the magnet and charge are unchanged. This cannot be. Something would have to supply additional field energy to compensate for the quantity absorbed by the detector. But (apart from the charge and magnet, whose contributions have already been taken into account) there is nothing in this example but empty space, which is not a source of energy. Therefore, even if energy is really circulating in this case, classical electromagnetic theory tells us that this circulation is undetectable (Rosser 1997: 320–1).

We can put a mark on a bit of solid matter (such as Lodge's suitcase) in order to follow its path. We can even mark a bit of fluid (with dye, for instance) in order to trace its career. But we cannot mark a bit of energy in order to verify that it has moved someplace else. Which parcels of energy have moved where appears to be a dangler. This counts to some degree (by a kind of argument we have seen before) against Lodge's view that "energy has identity like matter, and not mere conservation" (1885: 293).[11] If parcels of energy lack identities, there is no sense in which two energy parcels (5 joules each) *exchanged places* in the course of the "energy circulation" over the last few seconds. Rather, the situations before and after the "exchange" were *exactly the same*. Sir James Jeans (English physicist and astronomer, 1877–1946) presents this argument elegantly:

[T]he attempt to regard the flow of energy as a concrete stream always defeats itself. With a stream of water, we can say that a certain particle of water is now here, now there; with energy it is not so. The concept of energy flowing about through space is useful as a picture, but leads to absurdities and contradictions if we treat it as a reality. Professor Poynting gave a well-known formula which tells us how energy may be pictured as flowing in a certain way, but the picture is far too artificial to be treated as a reality; for instance, if an ordinary bar-magnet is electrified and left standing at rest, the formula pictures energy flowing endlessly round and round the magnet, rather like innumerable rings of children joining hands and dancing to all eternity round a maypole. The mathematician brings the whole problem back to reality by treating this flow of energy as a mere mathematical abstraction.   (1932: 129; see also 1960: 519)

The maypole – an analogy to rival Lodge's lost luggage! But let's allow Lodge to have the last analogy. Heaviside held this circulation of energy to be a "strange and apparently useless result" (useless since it cannot be measured and involves no change in the energy distribution) "serv[ing] to still further cast doubt upon the 'thingness' [!] of energy" (1925: II, 527). Lodge replied to his friend:

> The circulation of matter – for instance in the inner circle of the Metropolitan railway – is, I suppose, considered useful. The circulation of commodities is the essence of commerce. So does the circulation of energy constitute the activity of the material universe.   (1885: 293)

Picturesque, but surely beside the point. If energy is real stuff, then of course its flow (circular or otherwise) must also be real. However, since the alleged circulation cannot be detected, our observations cannot give us *directly* any reason to believe in it.

Here is a similar example not involving fields. Suppose billiard balls A, B, and C are all moving. Suppose A collides with B, increasing B's speed and diminishing A's. If bits of energy have continuing identities, then some of A's kinetic energy was transferred to B. Next, B collides with C, increasing C's speed and diminishing B's. Some of B's kinetic energy was conveyed to C. Was any of the kinetic energy transferred to C originally A's, or did all of it initially belong to B? The principle of energy conservation fails to tell us, and there is no way to mark a bit of energy in order to follow its path. To avoid hypothesizing that there is some hidden fact about which energy goes where – a "dangler" (like the electrostatic potential's absolute value at a given location)

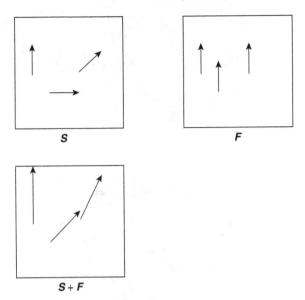

Figure 5.12   Adding an arbitrary constant **F** to **S** at various locations.

– we could agree with Jeans in interpreting energy as mere shorthand for a certain algebraic combination of qualities – in other words, as having no reality despite great usefulness by virtue of being a conserved quantity.

However, to avoid interpreting the Poynting flux over an open surface as the energy current crossing that surface, it is unnecessary to go so far as to deny that energy is some sort of stuff. The Poynting vector arose as a solution to the continuity equation, demonstrating that classical electromagnetic theory is consistent with energy's being a real, continuously flowing substance. But the Poynting vector is not the *only* solution to the continuity equation. Suppose that at every location, we add the same vector **F** to the Poynting vector **S** (figure 5.12). The result (**S** + **F**) is also a solution to the continuity equation, since for any region of space, **S** + **F** agrees perfectly with **S** regarding the rate of change in the quantity of field energy contained there (figure 5.13). That is, **S** and **S** + **F** agree on the energy flux over any *closed* surface, but they disagree on the energy flux across various *open* surfaces. Consequently, if we consider energy to be some sort of real stuff, then even though this stuff cannot flow across closed surfaces without flowing across open ones, there is a good argument for regarding **S**'s flux over a closed surface as the energy flow out of the enclosed region, but

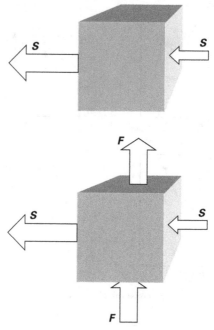

Figure 5.13   In the top figure, if **S** depicts energy flow, then more energy is flowing out of the cubical region through its left wall than into the region through its right wall; no energy is flowing through the region's other walls. In the bottom figure, **S** + **F** depicts energy flow, where **F** is uniformly upward. The net rate of change in the region's energy remains just as it was in the top figure, since the new flow of energy into the region through its bottom wall is perfectly balanced by the new flow of energy out of the region through its top wall.

remaining agnostic about whether **S**'s flux over any open surface represents the energy flow across that surface. Here is a statement of this attitude from J. J. Thomson:

> [Poynting's] interpretation of the expression for the variation in the energy seems open to question. . . . The problem of finding the way in which the energy is transmitted in a system whose mechanism is unknown seems to be an indeterminate one; thus, for example, if the energy inside a closed surface remains constant we cannot unless we know the mechanism of the system tell whether this is because there is no flow of energy either into or out of the surface, or because as much flows in as flows out. (1886: 151–2; also Abraham 1951: 131–2; Eyges 1972: 200–3; Robinson 1994)

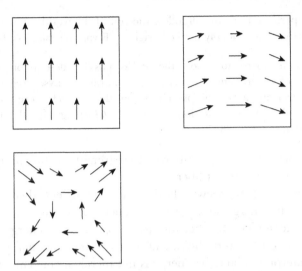

Figure 5.14   Three examples of divergence-free **F**s. Imagine them as the flow lines of water in a basin with no water entering or leaving. For any volume (a cube, for instance), the rate at which water is entering through some sides is exactly balanced by the rate at which water is leaving through the other sides.

This thoroughgoing agnosticism regarding the energy flux over open surfaces is distinct from the attitude displayed by my old textbook, which recommended interpreting **S**'s flux over *certain* open surfaces, but not others, as the actual energy flow.

Mathematically speaking, **S** + **F** satisfies the continuity equation if and only if **F** is "divergenceless"; that is, **F** can be anything as long as, over all closed surfaces, **F**'s flux is zero (figure 5.14). Despite these many solutions to the continuity equation, some have defended setting **F** = 0 (and thus interpreting **S**'s flux over every *open* surface as the actual energy flow) on the grounds of simplicity. For example:

> There are, in fact, an infinite number of different [solutions to the continuity equation], and so far no one has thought of an experimental way to tell which one is right! People have guessed that the simplest one is probably the correct one, but we must say that we do not know for certain. . . . Anyway, everyone always accepts the simple expressions we have found for electromagnetic energy and its flow. [Actually, not everyone; for just one example, see Jeffries (1992).] And although sometimes the results obtained from using them seem strange, nobody has ever found anything wrong with them – that is, no disagreement

with experiment. So we will follow the rest of the world – besides, we believe that it is probably perfectly right.   (Feynman et al. 1963: II, 27.6)

[O]ne is at liberty to choose the *simplest* possible description, in the absence of experimental results to the contrary. I myself regard this as a proper and appealing procedure, but this is much more a matter of taste than is the question of the . . . *total* energy [of the fields], etc.   (Romer 1966: 776–7)

Readers may judge for themselves the strength of this argument – whether we really are "at liberty."

Here is a different view: Classical electromagnetic theory fails to determine the energy flux across an open surface not because the theory is incomplete, but because *there is no such fact*: energy is not a stuff. On the other hand, the rate at which a volume's energy changes *is* to be interpreted literally; energy is not merely a convenient fiction. Rather, energy is a real, quantitative *property* over and above the various properties in the algebraic formula for it (m, v, q, and so forth). Not every property of an object consists of the object's possessing some sort of *stuff*. For example, to be happy is not to be filled with a large quantity of a special kind of stuff: "happiness." A body's velocity does not measure the amount of a stuff that it possesses. Likewise, neither a body's kinetic energy nor a field's energy is stuff. There are no energy parcels with continuing identities. Hence, strictly speaking, there is no such thing as energy's transfer, conversion from one form to another, or flow over a surface (open or closed). Therefore, we may disregard the alleged energy circulation around a magnet near a stationary charge: it corresponds to nothing in reality. There is no fact about whether S or $S + F$ (for some non-zero divergenceless F) is the correct depiction of energy flow, since they agree on the energy flux across a closed surface – that is, the rate at which the enclosed volume's energy is changing. Their disagreement regarding the energy flux over an open surface does not matter because, on this interpretation, there is no actual energy stuff the flow of which may be described accurately or inaccurately.

While not taking energy bits to have individual identities, this interpretation takes energy to be a real property. So energy has sufficient ontological status to underwrite the reality of fields, and hence spatiotemporal locality. Since energy is a property, any energy (like velocity) cannot exist without something possessing it. Thus, field energy requires a field.

This view appears to be a nice compromise. However, philosophy (unlike politics) does not operate through compromise.

7 A Moral Regarding the Testability of Theories

Our initial argument against retarded action at a distance was that it violates energy and momentum conservation: we need something to store energy and momentum during the interval between a cause at one charged body and its effect at another. We have been investigating whether fields store this energy and whether fields are therefore real (and thus serve as local causes of forces). In focusing on energy, I have said little about momentum, but analogous considerations apply. Just as $E^2/8\pi$ and $B^2/8\pi$ are the energy densities ascribed to the electric and magnetic fields (respectively), so the momentum density G ascribed to these fields (considered together) is $[1/4\pi c]\ [E \times B]$. (G is a vector quantity since momentum is a vector; "cross product" vector multiplication was defined in figure 2.6.) This field momentum exactly compensates for the non-conservation of the bodies' momentum in dynamic cases like the one we examined in section 2 (Page and Adams 1945). The total momentum in the electric and magnetic fields ($\int G\ dV$ over all space) plus the bodies' total momentum is a conserved quantity, just as the total electromagnetic field energy ($\int [E^2 + B^2]dV/8\pi$ over all space) plus the bodies' total kinetic energy is conserved (assuming that there is no change in the total quantity of all other forms of energy, such as gravitational potential energy).

We saw that as far as energy conservation was concerned, we could add an arbitrary quantity C to $E^2/8\pi$, so long as the total energy contributed by C ($\int C\ dV$ over all space) would remain constant no matter what. But we set $C = 0$ on the grounds that there can be no energy stored by the field at a location where there is no field (that is, where $E = 0$). Similarly, as far as momentum conservation is concerned, the momentum density could be $[1/4\pi c]\ [E \times B] + K$ for an arbitrary K, so long as the momentum that K contributes over all space would remain the same no matter what. But we constrain K to be zero if we make the sensible assumption that no momentum is stored by the fields where E and B are zero.

That a field can possess momentum, though it is not a body in motion, might strike you as bizarre. But it shouldn't, if you are already content with potential energy. Not all energy is kinetic, so by the same token, not all momentum is associated with bodies in motion. Both non-kinetic energy (known to its friends as "potential energy") and non-kinetic momentum (which, I guess, could be termed "potential momentum") might be interpreted as belonging to a field. That momentum can take different forms should seem no more bizarre than that energy can take different forms: kinetic energy, chemical-bond energy, electric potential energy, gravitational potential energy, and so forth. (Of course, it should seem no less bizarre either. The existence of different forms of energy returns us to the question of what they are all forms of: Is there a stuff that takes on different forms, just as water can be a solid, liquid, or gas? What makes the various "forms of energy" all forms of the same thing?)

Notice that $G = [1/4\pi c] [E \times B] = S/c^2$. Could this striking relation be mere coincidence? We asked similar questions in chapter 4, such as whether the correlation between the superior planets' moments of opposition and maximal brilliance could be coincidental. Suppose we concluded that this simple relation between G and S is no coincidence – that there must be some deep, undiscovered connection between momentum and energy in virtue of which the field momentum density times c^2 equals the field energy flux density. This conclusion would allow us to remove the indeterminateness in the energy flux density: Since S (the Poynting vector) equals $G\ c^2$, but $S + F$ (for any non-zero F) fails to equal $G\ c^2$, S must be the actual energy flux density.[12] In favoring S over $S + F$, this argument favors interpreting energy as stuff over interpreting energy as a property – since as a property, only the flux over a *closed* surface is meaningful, and this is the same for S as for $S + F$.

Of course, we cannot tell by direct experiment whether S or $S + F$ is the energy flux density, since they agree on the rate of energy flow into some region, which is all we can measure. But if we adopt the premise that energy flux density is related to momentum density, then we gain a new means of learning about the energy flux density. There is a general moral to be drawn here: No matter how far removed a certain fact seems to be from what we can directly observe, we cannot conclusively rule out the possibility that further research will yield well-confirmed theories that create a bridge between our observations and that fact, enabling it to be ascertained. Of course, that may never actually happen; humanity may fail to come up with a suitable bridging theory. It may not even be our fault: danglers may exist, which make no difference to anything, including our observations. But we can never know *for sure* that a certain hidden fact will always remain hidden. For example, we cannot foreclose the possibility that a fact that is dangling according to our current theories will someday be discovered not actually to dangle.

Perhaps a circulation of energy that leaves the energy distribution unchanged would be *undetectable*. But this does not mean that we cannot *know* whether or not there exists such a circulation. Whether we can infer S to be the actual energy flux density depends upon the premises that we can reason from. A newly justified premise – asserting a relation between the energy flux density and momentum density – may make available previously unforeseen inferential routes to the energy flux density. Of course, we would still like to know *why* such a simple relation exists between energy flux density and momentum density. (You will have to wait until chapter 8.) But we do not need to explain this relation in order to be justified in believing that it holds, any more than we need to know *why* the speed of light is 300,000,000 m/s before we can know *that* it is.

Certain philosophers and scientists sometimes say that various events are so distant in time or space, and various facts involve things so small or utterly undetectable, that observations could never, even in principle, tell us about these events or facts. But what an observation can reveal depends upon the background theories that we can bring to bear upon that observation. An observation could perhaps someday be shown to be relevant even to questions like "What, if anything, caused the Big Bang?" or "What, if anything, are quarks made of?" In 1835, Auguste Comte (French philosopher, 1798–1857) wrote regarding the stars:

> While we can conceive of the possibility of determining their shapes, their sizes, and their motions, we shall never be able by any means to study their chemical composition or even their density.   (1968: II, 2)

But by 1859, the theory of spectral analysis enabled Gustav Kirchhoff (German physicist, 1824–87) to infer from terrestrial observations that sodium exists in the Sun. (Of course, the Sun's chemical composition was never thought to be a dangler; it was considered merely too remote from what we can directly observe ever to be discovered.)

The simple relation between G and S may lead us to regard evidence that G correctly describes the field momentum density as evidence that S correctly describes the field energy flux density, even

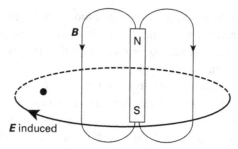

Figure 5.15 A short solenoid (with its north and south poles marked, and sample magnetic field lines shown) near a stationary charge (the dot). As the solenoid's **B** field is increased, an **E** field is induced.

when this requires that bits of energy be moving perpetually around a stationary magnet near a stationary charge (Furry 1969; Graham and Lahoz 1980). Although this case appears static, the field must possess momentum (in accordance with **G**) or else momentum conservation fails. To see a simple example, replace the bar magnet with a short solenoid, so that we may vary the magnet's strength by varying the current (figure 5.15). Before any current flows through the solenoid, both the charged body and the solenoid are at rest (and there is no magnetic field), so the system's total momentum is zero. Now suppose we slowly increase the current through the solenoid, thereby increasing its magnetic field. In chapter 2, I mentioned that outside a long solenoid, the field is negligible (except near the openings of the tube formed by the solenoid's windings). But here we have a short solenoid; its field is like a bar magnet's. As the solenoid's current is increased, its magnetic field at the charged body's location increases, and this changing magnetic field is accompanied by an electric field in accordance with the modern form of Faraday's law of electromagnetic induction: ***curl E*** $= -1/c$ ***B'*** (from the close of chapter 2). This electric field causes the charge q to feel a force q***E***. Suppose we balance this force by holding the charged body still with our hands. Once we have increased the solenoid's current to a certain level, suppose we hold it steady at that level. The solenoid's **B** is then no longer changing, and so the induced **E** disappears. We no longer need to apply any force to the charged body in order to hold it still. But we exerted a force on the charged body while the solenoid's current was increasing. As we saw at the start of section 4, a force **F** exerted from t_1 to t_2 changes the system's momentum by $\int$ ***F*** dt taken from t_1 to t_2. That additional momentum is not possessed by the solenoid or the charged body (since both have remained at rest). But momentum conservation is upheld if we ascribe momentum (with density G) to the field (Calkin 1966; also Feynman et al. 1963: II, pp.17-6 and 27-11; Pugh and Pugh 1967; Pugh and Pugh 1970: 245–9).

It is time to conclude this long chapter. As we have seen, the view that energy, momentum, and fields are real can survive many challenges, lending some support to spatiotemporal locality. But more challenges will come. Furthermore, although we have *rebutted* some arguments *against* believing energy, momentum, and fields to be real, we have not encountered a compelling argument *for* their reality. Instead of regarding $\int [E^2 + B^2] dV / 8\pi$ (taken over all space) as measuring the total energy stuff contained in the field, we might regard it merely as a quantity that can be added to various other terms (such as $(\frac{1}{2})mv^2$ for each body) to create a conserved quantity. Likewise, we might treat $\int G \, dV$ (taken over all space) as just a quantity that combines with various other terms (such as mv for each body) to make a conserved quantity. As mere calculational devices, energy and momentum do not have distributions in space. So there is no need for fields to be real in order to *contain* energy and momentum (recalling Maxwell's remark from the close of his *Treatise*). Classical physics fails to resolve our questions about energy's ontological status, the field's real-ity, and spatiotemporal locality. We must await Einstein's theory of relativity.

Discussion Questions

You might think about . . .

1   Maxwell offers this critique of a proposal involving retarded action at a distance:

> From [this proposal] we may draw the conclusions, first, that action and reaction are not always equal and opposite, and, second, that apparatus may be constructed to generate any amount of work from its resources.
>
> For let two oppositely electrified bodies A and B travel along the line joining them with equal velocities in the direction AB, then if either the

potential or the attraction of the bodies at a given time is that due to their position at some former time . . . , [then] *B*, the foremost body, will attract *A* forwards more than *A* attracts *B* backwards.

Now let *A* and *B* be kept asunder by a rigid rod.

The combined system, if set in motion in the direction *AB*, will pull in that direction with a force which may either continually augment the velocity, or be used as an inexhaustible source of energy.

I think that these remarkable deductions from [retarded action at a distance] theory can only be avoided by recognizing the action of a medium in electrical phenomena.   (1890: II, 137–8; also 1995: 355)

What is Maxwell's argument here against retarded action at a distance? Are energy and momentum conserved in the case he describes if it involves retarded action at a distance? How are these "remarkable" (in other words, preposterous) conclusions avoided if spatiotemporal locality holds? (Think carefully about the "ripples." What is their speed?)

2   In chapter 2 (section 2), I remarked that if the electric field's ontological status supports spatiotemporal locality, then we face the question: Does each charged body have its own electric field, extending to infinitely great distances (though diminishing there to negligible strength) and overlapping the electric fields of other charged bodies? Or is there only a single electric field extending everywhere, to which all charged bodies contribute? How does the fact that a field's energy density is $E^2/8\pi$ supply an argument for one of these options and against the other?

3   In 1894, Hertz considered it "an open question" whether energy is a stuff:

At the present time many distinguished physicists tend to attribute to energy so many of the properties of a substance that they assume that every smallest portion of it is associated at every instant with a given place in space, and that through all the changes of place and all the transformations of the energy into new forms, it retains its identity. . . . [P]otential energy . . . does not lend itself at all well to any definition that ascribes to it the properties of a substance. The amount of a substance is necessarily a positive quantity; but we never hesitate in assuming the potential energy contained in a system to be negative. When the amount of a substance is represented by an analytical expres-

sion, an additive constant in the expression has the same importance as
the rest; but in the expression for the potential energy of a system an
additive constant never has any meaning. Lastly, the amount of any
substance contained in a physical system can only depend on the state
of the system itself; but the amount of potential energy contained in a
given system depends on the presence of distant masses, which perhaps
have never had any direct influence on the system. . . . All these are
difficulties that must be removed or avoided by the desired definition of
energy.   (1994: 344–5; also Jungnickel and McCormmach 1986: II, 224–6)

What are Hertz's worries? How are they transformed if the field is real
and possesses energy?

4   A point body, with charge q, is all alone in the universe. Suppose
we calculate the system's electric potential energy by taking each pair
of point charges, multiplying the two charges, dividing by their separation, and adding all of the results (as in section 4). This obviously yields
zero in this example, since there is only a single point charge. On the
other hand, suppose we determine the field's energy by adding the
$(E^2/8\pi)$s for all regions of space. Clearly, E is not zero anywhere; it is
q/r^2. So this sum will not be zero. The two schemes give different
answers! How, then, can a system's electric potential energy be interpreted as the field's energy?

5   Consider this excerpt:

Consider two pendula, 1 and 2, that are coupled by a spring that joins
them. . . . The system can be started by holding pendulum 2 fixed in its
equilibrium position while displacing pendulum 1 and setting it in
motion. To accomplish this displacement and beginning motion, we
would have to apply a force to the pendulum 1, let it act over a certain
distance, and then release the bob. The *work* done on the bob, and
hence the *energy* given it in such a displacement, is of course calculable
as the integral of the net force on the bob and the displacement. Given
the initial conditions, the subsequent motion of this coupled system is
well known: Pendulum 1 has a certain initial amplitude of motion. After
a period of time, it will come to rest and pendulum 2 will be in motion
with the full amplitude that pendulum 1 had. Still later pendulum 1 will
regain this amplitude, and so it goes, periodically. . . . In [the layman's
language we have just used of "giving energy to the bob"], it is only
natural, when . . . 2 has attained the amplitude that 1 had and 1 is at
rest . . . , to say that energy . . . flowed from 1 to 2 [and to describe] the

behavior of the system in terms of a periodic *flow* of energy from one pendulum to the other.... However, our layman would encounter difficulties when he began to examine the concept of flow more closely, by analogy with the flow of charge or matter. What is the path the energy takes? Does it wind round and round along the helical coils of the spring? If so, given a second spring of some other material but the same spring constant, so that the length along the helix is different, why does the longer path of energy flow not manifest itself in a different behavior of the system?   (Eyges 1972: 201–2)

Is this a good argument?

6   Here is one sort of argument that has sometimes been given against the idea that energy is stuff that flows continuously from one location to another. Imagine a small motor mounted on a board and powered by a battery. By means of a belt (that is, a strap), the motor drives a paddlewheel that stirs the water in a basin mounted on the board (figure 5.16). Hertz argues:

In ordinary language, we say – and no exception need be taken to such a mode of expression – that the energy is transferred from the [battery and motor] by means of the strap to the [paddlewheel], and from this again to the [water]. But is there any clear physical meaning in asserting that the energy travels from point to point along the stretched strap in a direction opposite to that in which the strap itself moves? And if not, can there be any more clear meaning in saying that the energy travels from point to point along the wires, or – as Poynting says – in the space between the wires? There are difficulties here which badly need clearing up.   (1893: 277 – see also 200, 220; also Buchwald 1985: 41–3)

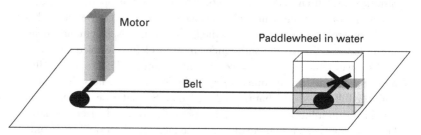

Figure 5.16   The apparatus for examining the concept of energy flow; the motor turns the belt, which turns the paddlewheel in the water.

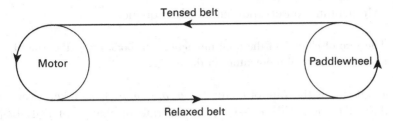

Figure 5.17 A schematic diagram of the apparatus in figure 5.16; arrows indicate the direction in which the belt moves. (After Buchwald 1985: 42.)

In other words, if energy flows continuously (rather than hops across the gap between the motor and the paddlewheel), then surely it must travel via the strap. Indeed, the motor pulls on the upper half of the strap (figure 5.17), and consequently (like a succession of springs that have all been stretched), this part of the strap contains potential energy; it is under tension: taut. However, if this strap carries potential energy, surely it carries that energy *away* from the paddlewheel, since this is the direction in which the strap is moving. Each little stretched spring (with its bit of potential energy), linked to others that together comprise the tensed belt, is moving to the left (in figure 5.17), carrying its potential energy leftward, whereas the energy is flowing to the right! In Bridgman's words,

> [T]he energy in the belt is running the "wrong" way, against the motion of the belt from motor to flywheel.... [This] may start us to wondering whether after all we should be so pleased with our treatment of mechanical energy as a thing.   (1941: 36, also 33)

(The slack half of the belt is traveling *toward* the paddlewheel from the motor, but surely *it* is not conveying energy, since it is no different in its tension from what it would be if the motor were turned off.)

Is this a good argument? You might wish:

(i)   to think hard about the tensed belt as a succession of linked stretched springs (Where do the *stretches* begin when the motor starts to turn?);

(ii)   to compare this example to Poynting's treatment of a current-carrying wire;

(iii) to consider what fundamental force produces tension in a belt (or spring).

7   Momentum conservation entails this equation:

> The rate of change of the total momentum in bodies = − (the rate of change of the total momentum in the field).

The left side is the rate of change of $m_1\boldsymbol{v}_1 + m_2\boldsymbol{v}_2 + \ldots$ taken over all bodies. The right side is (−1) times the rate of change of $\int [1/4\pi c]$ $[\boldsymbol{E} \times \boldsymbol{B}]\ dV$ taken over all space. A careful interpreter of classical electromagnetic theory has written:

> It is assumed nowadays that changes in the electromagnetic interaction between moving charges take a finite time to propagate at the speed of light. It is stretching the imagination to interpret [the above equation] by assuming that the rates of change of the fields $\boldsymbol{E}$ and $\boldsymbol{B}$ at a point well away from any of the moving charges in the system can contribute in a causal way to the rates of change of the linear momenta of distant charges at that precise instant. It is best to interpret [the above equation] as a book-keeping exercise carried out at a fixed time rather than as a causal relation between the fields and particles at a particular instant of time.   (Rosser 1997: 310)

Do you agree? How did Poynting deal with the analogous question for energy? What about a principle of the continuity of momentum?

8   Regarding the interpretation of electric potential energy as belonging to the field, Pilley correctly remarks:

> If we were consistent in adopting such an explanation we should also have to take the view that the energy expended by a man in climbing a ladder was somehow stored up in the space through which he had passed [that is, by the gravitational field there] and reappeared [in the man] if he fell or jumped off of it. Actually it has never been the fashion to explain gravitational action in this way.   (1933: 96)

Is this merely a "fashion" or is some important difference between classical electromagnetism and Newtonian gravity responsible for it?

Notes

1   In this case, neither body exerts a magnetic force on the other. The magnetic field lines curve around a current, so there is no field along the

axis of the current. Since each particle lies along the axis of the other's motion, neither evokes a magnetic field at the other's location. Hence, there are no magnetic forces that need to be added to the calculation.

2   However, Newton's third law ("Every action is accompanied by an equal and opposite reaction") is still violated, even if fields are real. Bodies do not exert forces on fields; bodies alone feel forces. Newton's third law was thus abandoned before relativity theory came on the scene.

3   To be more historically accurate, Maxwell hypothesized that the repository of this energy is the aether (a kind of matter) in which the field resides, the field being a state of the aether; see chapters 2 and 7. For Maxwell, energy can exist only in matter because matter is defined dynamically: as whatever carries energy and momentum (Harman 1998: 190–1).

4   This relation between momentum and energy will become especially important in chapter 8, when we see how relativity theory breaks down the distinction between time and space, and hence between momentum and energy, thereby unifying them.

5   It is not possible to define a potential energy function for every type of force – for the force of friction, for example, where the work depends not just on the initial and final positions, but also on the path taken to get there. But I exclude such non-fundamental forces from consideration here.

6   You might wonder where that minus sign came from in the penultimate step. Notice that I have been sloppy in my use of "r." In connection with Coulomb's law, I have been using "r" for the distance from q_1 to q_2, so that r gets *smaller* as we move q_2 towards q_1. But in connection with the integral taken along the path traversed by q_2, I have been using "r" to denote the distance covered by q_2, so that as we move q_2 towards q_1, it has traversed more of its path, and so r gets *larger*. In terms of "r" in Coulomb's law, each element of the path *towards* q_1 is a small *diminution* of r: $-dr$.

7   Valence numbers are a simplified account of chemical bonding: For instance, hydrogen peroxide (H_2O_2) is also a stable compound. But valence numbers can still illustrate my point; they could have provided a perfectly accurate theory, and my concern is the interpretation of theories.

8   I shall have more to say about "additivity" in chapter 8. Of course, a definite quantity of *kinetic* energy seems to belong to the cart. Its own m and v determine its kinetic energy, and so it seems easy to envision the cart's kinetic energy as located wherever the cart is. In this respect, kinetic energy appears to stand in stark contrast to potential energy. But they are actually similar in this regard, since the kinetic energy's value depends upon the zero state of motion. We'll look into this further in chapter 8.

9   The minus sign arises from the fact that to move the sphere's part *inward* requires that we *diminish r*. See note 6.

10 If a region contains a point charge, then this integral becomes infinite. If the charge itself is taken not to contribute to E at $r = 0$, then the resulting E^2 cannot be integrated. The problem of deriving useful results concerning the field's energy when there are point charges can be finessed in various ways (Boyer 1971; Cole 1999; Dirac 1938; Rohrlich 1965). This issue fails to arise if every charged body has finite non-zero volume.

11 By the way, Maxwell wrote: "We cannot identify a particular portion of energy, or trace it through its transformations. It has no individual existence, such as that which we attribute to particular portions of matter" (1892: 166). But Maxwell died in 1879 (the year of Einstein's birth!), five years before Poynting discovered his vector.

12 I am unaware of any nineteenth-century physicist making this argument. There is at least one article (Birkeland 1894) in which a physicist imposed various constraints on the energy flux density in order to render it unique (see Romer 1982: 1168). But Birkeland's constraints strike me as *ad hoc*. (Kristian Birkeland (1867–1917) was the Norwegian physicist who explained how the aurora borealis is created by the Earth's magnetic field capturing charged particles emitted by the Sun.)

6
Is there Nothing but Fields?

1 Is Electric Charge Real?

Suppose that electric interactions obey spatiotemporal locality because the electric field is a real entity. Its state at a given place and time is caused ultimately by bodies at distant locations at earlier moments. A body's having a certain charge (and so forth) at a certain moment causes an event – a contribution to the field – that spreads out from that body like a ripple in a pond. At any location, the contributions from all the charged bodies sum to give the electric field's strength and direction there.

On this view, a body's electric charge is one of its fundamental properties. However, there is another interpretation that allows spatiotemporal locality to hold but does *not* take electric charge to be fundamental. Rather, facts about electric charge are nothing but facts about the electric field!

The key to this proposed reduction is Gauss's law (one of Maxwell's equations).[1] Here is one way to express this law. Consider $\boldsymbol{E}$'s flux through some closed surface – that is, $\int \boldsymbol{E} \cdot \boldsymbol{dA}$ over that surface. (Recall from chapter 5 that this is what you get when you divide the surface into small patches, multiply the area of each patch by $\boldsymbol{E}$'s component there perpendicular to the patch, add the results for the entire surface, and then take the value that this sum approaches as the patches become arbitrarily tiny.) Gauss's law says that $\boldsymbol{E}$'s flux over any closed surface equals 4π times the net charge q in the volume enclosed by the surface:

$$\int \boldsymbol{E} \cdot \boldsymbol{dA} = 4\pi q.$$

This law holds in all cases – regardless of how the surface is shaped, how the enclosed charged bodies are moving, how they are distributed within the region enclosed, and even whether the net enclosed charge (say, 20 statcoulombs) consists of two bodies of 10 statcoulombs each, or one body of 40 statcoulombs and another of –20 statcoulombs.

Gauss's law is often a convenient way of calculating the electric field if the distribution of electric charge is already known. But we are interested not in its practical uses but in the metaphysical possibilities it makes available. In particular, Gauss's law suggests the possibility of regarding the electric charge in a given region as merely something about the electric field on the surface enclosing the region. On this view, what makes it true that a certain region contains, say, 20 statcoulombs of charge is not certain properties possessed by the bodies in the region (namely, their electric charges). Rather, "The region contains 20 statcoulombs of charge" is made true by the fact that E's flux over the surface enclosing exactly this region, divided by 4π, is 20 statcoulombs. Remarks concerning electric charges are *literally* false (since there is no such thing as "electric charges" for those remarks to describe correctly). But interpreted *non-literally*, these remarks are made true (or false) by features of the electric field.

What about the charge at a point? Consider E's flux over the surface enclosing a region that includes the point. Allow the surface to shrink gradually around the point, like a balloon deflating. E's flux over the surface approaches some value as the surface contracts arbitrarily tightly around the point. That value, divided by 4π, equals the charge at the point, according to Gauss's law. (If there are no point charges, Gauss's law can still be applied to point-sized "regions": E's flux over the surface divided by the volume of the region enclosed approaches $4\pi\rho$ in the limit, where ρ is the "charge density" at the point – see box 2.1.) Claims that literally concern charge at a given point are thus reinterpreted as describing the electric field in the point's neighborhood.

On this view, the field's state here now is not caused by the past locations of various charges (understood literally) since on this interpretation, talk of "charges" is true only insofar it is understood non-literally: as a means of describing the field. So what *does* cause the field's state here now? The answer is: the field's conditions at other places and earlier times (which we might, for convenience, describe in the language of "charges"). In other words, on the original view, according to which charges are real, a field's strength and direction here now is given by events (the ripples) that earlier began spreading out from bodies as a result of their charges then. On the new proposal, there are no charges to cause these ripples, and so there is no reason to retain these ripply events as the means by which charges cause remote forces without violating spatiotemporal identity. Instead, a complete cause of a field's strength and direction here now consists simply of the field's condition (its strength and direction, and how these were changing) between now and t seconds ago at all points between here and ct centimeters away from here. A complete set of causes can thus be found within any span of time and space around here now, no matter how small – as spatiotemporal locality demands.

Consider a body feeling an electric force. On the original view, according to which charges are real, this force F is caused by the body's charge q and the electric field E at the body's location: $F = qE$ (equation 2.5). However, if talk of charges is really just shorthand for certain features of E, then the force on the body is caused by E at the body's location and E at the points on surfaces that surround the body arbitrarily closely. Spatial locality is obeyed since for any finite nonzero value, the distance between the body (where the force occurs) and the enclosing surfaces can be reduced beyond that value.

On this view, then, electric fields act on bodies, but a body's charge is nothing over and above a feature of the electric field. But can we go even further – can we regard bodies themselves as features of fields?

2 Faraday's Picture

In chapter 2, we saw that if fields are real, then we can avoid the problems that arose in chapter 1 when we tried to understand what happens when two bodies collide. (Do they overlap at a point in space? How is that possible – surely, two bits of matter cannot be at the same place at the same time, can they? Do the colliding bodies have a gap of one point between them? Then, apparently, they are not really touching!) Perhaps each body is surrounded by a repulsive field that is negligible except at very small distances, where it becomes very great. When two bodies come near enough, each body is acted upon by the other's field, strongly forcing the bodies apart. We call this a "collision" though the two bodies never actually touch. Unlike two bits of matter, a field and a bit of matter can simultaneously have the same location without raising any difficulties. (The particles composing a macroscopic body must be far enough apart that their mutual repulsion is small, and so is overwhelmed by other forces attracting them to one another.)

On this view, a body is a real thing (either having some finite size or occupying a mere point) that is surrounded by a repulsive field, another real thing. But perhaps we can dispense with the body – eliminating it from our picture of what there is, leaving only the field. Our reason for believing in matter (that is to say: in hard, impenetrable, solid stuff) has *disappeared*, since collisions are being explained not by the hardness of the matter out of which bodies are made, but by the repulsive fields surrounding them. In an unpublished note to himself, we can watch Faraday exploring this line of reasoning:

> I press my finger against a piece of glass, and, because my finger is resisted by it I say it is *hard*; but how does this hardness or resistance arise? by a *force of repulsion* which existing in the particles of the glass & in the particles of my finger prevents their coming nearer to each other than a certain distance, fixed for the circumstances but varying if the circumstances vary. I say again that the glass is *hard* because its particles resist displacement; not that they are touching . . . for we can easily place the particles nearer to, or farther apart from, each other by pressure, heat, &c . . .
>
> All the properties, therefore, by which we are made aware of the presence of, & recognise, the matter are dependent on *forces* acting at some distance from the real nucleus [the particle of matter]; and of that, as a thing by itself, we cannot in any way be conscious. So then, for aught we can tell, the supposed material nuclei, instead of being so large as nearly to touch each other, may be of only half that size or diameter, or less still, or even mere points; for whether these nuclei be of the larger size, or of the next size, or of the still smaller third size, or little more than a mere point, if they have a constant amount of power [that is, power to repel one another via their fields] for all the sizes, their effects and properties will be the same. . . .
>
> Well, then; – as we cannot recognize this nucleus by any property or force it has, independant [*sic*] of those which are shewn by all the phenomena of nature to act at a distance from it, what reason is there to suppose that it exists at all? (Levere 1968: 105–7; italics in original)[2]

Faraday concluded that we have no reason to suppose that there are particles of matter, little hard bodies that exist over and above fields of force. Accordingly, he publicly proposed (1844) that the world contains no matter at all in this sense, only fields.[3]

Before Faraday, a few other natural philosophers had also thought it unreasonable to explain collisions between macroscopic bodies (like billiard balls) by appealing to the hardness of the fundamental particles composing those bodies. For example, William Whewell (English historian and philosopher of science, 1794–1866) wrote:

> According to the Atomic Theory . . . , the properties of [macroscopic] bodies depend upon the attractions and repulsions of the [fundamental] particles. Therefore, among other properties of [macroscopic] bodies, their hardness depends upon such forces. But if the hardness *of the [macroscopic] bodies* depends upon the forces, the repulsion, for instance, of the particles, upon what does the hardness *of the particles* depend? What progress do we make in explaining the properties of [macroscopic] bodies, when we assume the same properties in our explanation? and to what purpose do we assume that the particles *are* hard?   (1967: I, 432; italics in original)

In terms of fields, we might put Whewell's argument this way: if we are not content to take the hardness of macroscopic bodies as an unexplained explainer, but rather wish to account for their apparent hardness in terms of repulsive fields surrounding their constituents, then by the same token, we should not be content to take the hardness of those constituents as a brute fact about them. We should instead explain their apparent hardness, in turn, as the result of repulsive fields. Their matter's alleged hardness, then, has no explanatory work to do. (If we are willing to take the hardness of constituent particles as brute facts about them, then we might just as well have taken the hardness of macroscopic bodies as unexplained explainers in the first place and done without the constituent particles.) Whewell thinks that there is no evidence that bits of hard matter constitute macroscopic bodies because the supposed hardness of those bits is not used to explain any of our observations.

Whewell would presumably be happier with Faraday's explanation, which sees "bodies" not as hard matter with repulsive fields surrounding them, but simply as fields of repulsive force.[4] On this view, those fields are not *caused* by the matter they surround. Rather, the fields *are* the bodies; there aren't two kinds of things, just one. A claim about some elementary particle's location is made true (or false) by the location of the center of some field of repulsive force. A particle's *mass* refers not to the amount of matter (some kind of hard stuff) that the particle is made out of, but rather to the particle's inertia: its inertness, the degree to which it resists being accelerated by a force on it. Newton's second law, $F = ma$, says that the acceleration a that a body undergoes, in response to a force F, is less insofar as its mass m is greater. Of course, on Faraday's view, the force is felt not by a bit of matter, but instead by a repulsive field (possessing inertia) surrounding some center.

The repulsive field constituting a body is especially strong in a certain small region, but presumably, it behaves like gravity and electromagnetism in extending to arbitrarily great distances (though it may become arbitrarily small there). So on Faraday's picture, a body may be "centered" at a given location but is not confined to a nearby region. Rather, it extends to infinity. Thus, all bodies interpenetrate; that a given body is located in a given region does not exclude other bodies from that region. According to this picture:

> In gases the atoms touch each other just as truly as in solids. In this respect the atoms of water touch each other whether that substance be in the form of ice, water or steam; no mere intervening space is present. Doubtless the centres of force vary in their distance one from another, but that which is truly the matter of one atom touches the matter of its neighbors. . . .
> In that view matter is not merely mutually penetrable, but each atom extends, so to say, throughout the whole of the solar system, yet always retaining its own centre of force.   (Faraday 1844: 291, 293)

Of course, where one body's repulsive field is great, it takes more effort to force the center of another body's repulsive field into that region at the same time. Nevertheless, since bodies do not consist of impenetrably hard particles (whether of finite size or points), the centers of two bodies' repulsive fields could be made to approach arbitrarily closely or even to coincide, were enough force applied. (Faraday (1844: 292-3) speculated that this might happen, for instance, when various atoms combine to form one molecule.)

On Faraday's view, the "center" of a repulsive field (what Faraday calls a center "of power" or "of force"[5]) is not in the geometric center of anything – that is to say, equidistant from the edge in all directions – since the field surrounding it does not have an edge. Furthermore, the field need not be spherically symmetric. In other words, the field's strength may differ at the same distance in different directions from the "center" (Faraday 1844: 292). Presumably, the "center" should be understood to be where the repulsive field's strength reaches its maximum.

If each body is located everywhere at once, then all bodies are everywhere in contact, and so it would appear to be much easier for spatial locality to be satisfied. Nevertheless, spatial locality is not *automatically* satisfied. An *event* may be confined to a certain finite region of space even if a body is not, as when the event is a force's occurring at the center of a repulsive field (the "location" of the body's inertia).

So having used Gauss's law to eliminate charge from our picture of the universe's basic ingredients, we could follow Faraday and eliminate the hard particles, replacing them with fields centered on points endowed with inertia and capable (in principle) of coinciding – even of passing through each other. Faraday's view that fields alone are the fundamental real things is one of the earliest moves in what has now become a tradition of trying to *unify* fields and matter. (I shall say more about what "unification" means in the next chapter.) During the period just before the special theory of relativity was developed by Einstein (in 1905), several physicists (equipped with Maxwell's equations, as Faraday was not) tried to work out the idea that bodies were nothing more than local concentrations of the electromagnetic field. (Among the proponents of an "electromagnetic theory of matter" were the German physicists Max Abraham, 1875–1922; Wilhelm Wien, 1864–1928; and Gustav Mie, 1868–1957.) Later, Einstein also tried to work out a "unified field theory" according to which bodies would become identified with local concentrations of some generalized field. (On one interpretation of the general theory of relativity, space and time are ontologically subordinate to the gravitational field – Cao 1997.) Finally, in quantum field theory, the distinction between particles and fields seems to disappear, or at least to become much more difficult to draw (Redhead 1982; Teller 1995). But this topic falls outside the scope of what can be treated here.

How plausible is Faraday's picture as an interpretation of classical physics? I can think of some considerations possibly counting in its favor and also some possible objections to it. But I will present these as questions for you to ponder.

Discussion Questions

You might think about . . .

1   In section 1, I explained how Gauss's law opens the possibility of regarding charges as features of the field rather than as causes of the field, but I gave no reason why this view is better than its rival. It

might be argued that this view is more economical or unified in that it does not require one kind of thing to make claims about charges true and another to make claims about the electric field true. How strong is this argument? When is greater economy or unity of this sort a good reason for preferring one interpretation over another? Likewise, on Faraday's view, there aren't two kinds of things (fields and matter), just one. Does this make his view more likely to be true?

2   The idea that there are no solid things to which fields are attached – that it is "fields all the way down" – may seem bizarre. Concerning a view similar to Faraday's, John Mason Good (English physician and scholar, 1764–1827) wrote:

> [W]e are told that these points [the field centers] are endowed with certain powers; as those, for example, of attraction and repulsion. But powers must be the powers of something: what is this something to which these powers are thus said to appertain? If the ultimate and inextended points before us have nothing but these powers, and be nothing but these powers, then are such powers powers of nothing . . . ?   (1837: 45)

Try to express as clearly as you can just what basic presuppositions might prompt these feelings of unease with Faraday's view. Faraday himself seemed to try to make explicit the underlying source of these difficulties:

> Is the lingering notion which remains in the minds of some, really a thought, that God could not just as easily by his word speak power into existence around centers, as he could first create nuclei & then clothe them with power?   (Levere 1968: 107)

Can you bring yourself to sympathize with this objection? Or is it based on a fundamental mistake or some kind of misunderstanding of Faraday's picture?

3   Faraday regarded a body as extending over all regions where its force could be felt. This force includes not only whatever repulsive force is responsible for what we ordinarily take to be matter's solidity, but also gravitational, electric, magnetic, and any other forces. Does the fact that the "centers" of these various, entirely distinct fields *coincide* lend any support to the notion that particles exist as entities over

and above fields? Do we need an explanation of their coinciding? Can we imagine a universe in which these centers do not coincide?

4   The centers of repulsive fields are stipulated by Faraday to possess inertia, as a brute matter of fact. Maxwell objects to Faraday's view on these grounds:

> It is probable that many qualities of bodies might be explained on this supposition, but no arrangement of centres of force, however complic- ated, could account for the fact that a body requires a certain force to produce in it a certain change of motion, which fact we express by saying that the body has a certain measurable mass. No part of this mass can be due to the existence of the supposed centres of force. (1908: 86)

A similar point has been made by Emile Meyerson (French philoso- pher of science, 1859–1933):

> Let us penetrate the forces which surround this atom. The centre of these forces is a point which is, properly speaking, empty. . . . How can this *nothing* resist motion? How, once in motion, can it preserve it? How, in a word, can it possess a mass, manifest inertia?   (1962: 76)

Is this a powerful argument against Faraday's picture?

Notes

1   For a historically more accurate account of how Maxwell and his fol- lowers interpreted electric charge, see Buchwald (1985: 23–34).
2   Faraday's views regarding action at a distance were subtle and changed over time; see Gooding (1978) and Levere (1971: 68–106).
3   Beware: some commentators interpret Faraday as taking fields to be dispositional properties. On this interpretation, Faraday's conclusion that the world contains no matter at all, only fields, is the thesis that disposi- tions (such as the impenetrability of a "body") have no categorical bases; it is dispositions all the way down. However, I believe this interpretation of Faraday to be incorrect. Lines of forces, for him, were things possessing categorical properties. (That they are known and identified only through their powers raises issues we grappled with at the close of chapter 3.) The same attitude as Faraday's is evident in Hertz's remark quoted in chapter 3, near the start of section 3.

4   Faraday's ideas were anticipated in some respects by Roger Boscovich (Croatian natural philosopher, 1711–87) and Immanuel Kant (German philosopher, 1724–1804) – see Hesse (1965: 163–6 and 169–80) and Heilbron (1982: 58–9). But unlike Faraday, Kant may have contended that it is dispositions all the way down (see note 3).

5   For more on Faraday's use of the term "force," see Levere (1968: 104).

7

Relativity and the Unification of Electricity and Magnetism

1 Unification in Physics

In chapter 5, we saw how classical electromagnetic theory could be interpreted as describing real electric and magnetic fields, thereby supporting spatiotemporal locality. We wrestled with various issues arising from this interpretation, such as the ontological status of energy and the reality of the energy flow purportedly described by the Poynting vector. As we will see in chapter 8, Einstein's theory of relativity (published in 1905, 32 years after Maxwell's *Treatise on Electricity and Magnetism*) has a great deal to say about energy's ontological status and the reality of fields. This chapter will prepare us for that discussion. I focus here on just one of the consequences of Einstein's theory: that the electric field and magnetic field are not separate entities, but rather two sides of the same coin – the electromagnetic field.

We are all familiar with the fact that from different perspectives, the same entity can look different. For example, a penny seen from above looks circular, whereas seen from the side, it looks elliptical. The penny's appearance to us depends not just on the way the penny really is, but also on the vantage point from which we are gazing at it. In a roughly analogous way, Einstein's theory reveals that if electric and magnetic interactions obey spatiotemporal locality, then various combinations of electric and magnetic fields, as seen by different observers, are really the same single object (the electromagnetic field) appearing differently from different vantage points.

In other words, Einstein's theory *unifies* electricity and magnetism. If a given force is electromagnetic, there is no fact about whether it is caused by the electric field or the magnetic field. These turn out not to be rival possibilities. It is as if we were asked what kind of force causes the ocean tides: the kind of force that causes bodies to fall to Earth, or the kind of force that causes the planets to orbit the Sun. The answer, of course, is that these are actually the same force: gravity.

Classical (that is, pre-relativistic) electromagnetic theory reveals electricity and magnetism to be closely related in several respects. By examining these connections first, we will better appreciate the special sort of unity that relativity adds.

Let us begin with the most superficial connection that classical electromagnetism establishes. Electric and magnetic forces, along with forces of all other kinds (such as gravity), affect the motion of the body feeling them in accordance with Newton's second law of motion: $F = ma$ (where F is the net force on the body, m is the body's mass, and a is its acceleration). In other words, electric and magnetic forces are just two among the many kinds of forces, all of which are (in a sense) "unified" under Newton's second law.

Furthermore, as we saw in chapter 5, classical electromagnetic theory (as Maxwell interpreted it) associates a certain energy density with the electric and magnetic fields at each point in space. This energy can be transformed into or derived from heat energy, chemical energy (as stored in a battery), or energy of any other form. Although Maxwell wanted some parts of his theory (such as the electric potential) to be understood as mere convenient devices,[1]

> In speaking of the Energy of the field, however, I wish to be understood literally. All energy is the same as mechanical energy, whether it exists in the form of motion or in that of elasticity, or in any other form. The energy in electromagnetic phenomena is mechanical energy. (1890: I, 564)

The energy ascribed to the electric and magnetic fields is one form of the very same thing other forms of which are associated with chemical, thermal, elastic, kinetic, and gravitational phenomena, for example.

There is, then, a sense in which energy conservation "unifies" electricity and magnetism with other parts of nature:

> [T]he doctrine of the Conservation of Energy is the one generalized statement which is found to be consistent with fact, not in one physical science only, but in all. When once apprehended it furnishes to the physical inquirer a principle on which he may hang every known law

relating to physical actions, and by which he may be put in the way to discover the relations of such actions in new branches of science. . . .

The discussions of the various forms of energy – gravitational, electromagnetic, molecular, thermal, &c. – with the conditions of the transference of energy from one form to another, and the constant dissipation of the energy available for producing work, constitutes the whole of physical science, in so far as it has been developed in the dynamical form under the various designations of Astronomy, Electricity, Magnetism, Optics, Theory of the Physical States of Bodies, Thermo-dynamics, and Chemistry.   (Maxwell 1892: 102, 168–9; see Morrison 2000: 104)

A system's energy can be calculated even if a detailed mechanical model of the system is unavailable (at least as yet). The system may be a hot gas, for instance, where the forces between the individual gas particles (not to mention their various positions and velocities) are unknown. Or the system may include the medium between charged bodies whose local action on each body produces the electric or magnetic forces on it. Thus, energy conservation could in a sense "unify" a broader range of phenomena than even Newton's second law. Since Maxwell's arguments for his equations governing charges, currents, and fields do not depend on any specific mechanical model of the aether, his equations survived the demise of his own and every other proposal for understanding the electric and magnetic fields as the properties of some material medium.

As explained in chapter 5, classical electromagnetic theory takes the Poynting vector $\boldsymbol{S} = (c/4\pi)\boldsymbol{E} \times \boldsymbol{B}$ to be the energy flux density. $\boldsymbol{S}$ is a solution to the "continuity equation" for energy. That equation treats energy as if it were stuff that flows continuously through space. The first "continuity equation" was developed by Leonhard Euler (Swiss mathematician, astronomer, and physicist, 1707–83) for the flow of an actual fluid, such as water. In the 1820s, Joseph Fourier (French mathematician and physicist, 1768–1830) applied an analogous equation to heat flow. In other words, he treated heat as an incompressible fluid the total amount of which flowing into a certain volume equals the increase in the heat contained in that volume. That increase is indicated by the rise in temperature of the volume's matter (which stores heat). The rate at which heat flows across the surface enclosing the volume depends on the temperature difference between the enclosed volume and the region outside of it, just as the rate at which an actual fluid flows through a pipe depends on the pressure difference between the pipe's two ends.

This analogy between hydrodynamics and thermodynamics does not presuppose that heat really *is* some kind of *stuff* that flows. (Maxwell helped to develop the conception of heat we use today, according to which heat is merely random molecular motion.) The analogy between hydrodynamics and thermodynamics presupposes only that for certain purposes, heat may successfully be treated as some kind of flowing stuff. The connection between heat and fluid flow involves mathematics alone: equations of the same form (such as continuity equations) govern both phenomena (Maxwell 1990: 356; Morrison 2000: 65–6).

The mathematical analogy "unifying" hydrodynamics and thermodynamics was extended to electrodynamics by William Thomson (later Lord Kelvin – Scottish mathematician and physicist, 1824–1907). In the early 1840s (when he was still an undergraduate – isn't that intimidating!), Thomson showed that the electric potential is analogous to temperature. Just as heat flows at any point from hot to cold (specifically, in the direction of the greatest decrease in temperature), so a positive charge at a given point feels an electric force in the direction of the greatest decrease of electric potential (Wise 1981). In other words, Faraday's lines of force are analogous to lines of heat flow. By drawing this mathematical analogy, Thomson was able to use already-established results regarding heat conduction to arrive at entirely new results regarding electrostatics. As Thomson put it:

> Corresponding to every problem relative to the distribution of electricity on conductors, or to forces of attraction and repulsion exercised by electrified bodies, there is a problem in the uniform motion of heat which presents the same analytical conditions, and which, therefore, considered mathematically, is the same problem. . . . The problem of *distributing sources of heat* . . . is mathematically identical with the problem of distributing *electricity* [positive electric charges] *in equilibrium*. . . . In the case of heat, the *permanent temperature* at any point replaces the *potential* at the corresponding point in the electrical system, and consequently the *resultant flux of heat* replaces the *resultant attraction* of the electrified bodies, in direction and magnitude. The problem in each case is determinate, and we may therefore employ the elementary principles of one theory, as theorems, relative to the other.   (1872: 27–9; see Maxwell 1990: I, 356; Knudsen 1985: 153; italics in original)

Again, this analogy does not presuppose that electricity really *is* a fluid; Thomson said that he extended the hydrodynamic and thermodynamic mathematical machinery to electric phenomena "without involving even the idea of a hypothesis regarding the nature of electricity" (1872: 48; see Gooding 1980: 108–11; Moyer 1978). Later, Thomson broadened the analogy to cover magnetism as well (Harman 1998: 80, 88; Knudsen 1995: 72–3). This analogy led, in turn, to Maxwell's development of Faraday's lines of force in terms of the streamlines in the flow of an imaginary fluid (see box 2.3). Finally, of course, Poynting applied the same formalism to the flow of energy through electric and magnetic fields.

The fact that water, heat, and electromagnetic energy all flow in accordance with continuity equations suggested further discoveries and saved scientists from having to solve essentially the same problem repeatedly (Maxwell 1890: II, 258; 1990: 382). However, although the relation between the electric field and electric potential is the same as the relation between an unequally heated body and temperature, which in turn is the same as the relation between an unequally distributed material fluid (such as water) and pressure, the *entities standing in these relations* are obviously different in each case: an electric field is not actually made of heat or water! The "unification" achieved by drawing *physical analogies* (Maxwell 1890: I, 156; 1990: 376) among various processes does not reveal them all to be the same process variously disguised.

Classical electromagnetic theory identifies two other, especially intimate connections between electricity and magnetism. First, a body's electric charge q appears not only in the equation $F_e = q E$ (equation 2.5) relating the electric force F_e on the body to the electric field E at its location, but also in the equation $F_m = (q/c) v \times B$ (equation 2.7) relating the magnetic force F_m on the body to its velocity v and the magnetic field B at its location. Second, a changing magnetic field is always accompanied by a non-zero electric field (as dictated by Faraday's law of electromagnetic induction) and a changing E in empty space is always accompanied by a non-zero B. So classical electromagnetic theory portrays the states of the electric and magnetic fields as tightly interlinked.

However, classical electromagnetic theory fails to *unify* electricity and magnetism *in the deepest sense*: it does not reveal them to be one and the same thing, only appearing different to us as a result of what our vantage point happens to be (Maudlin 1996: 131–2; Morrison 2000: 107–47). Rather, classical electromagnetic theory depicts the electric and magnetic fields as two separate entities – though they are similar in certain respects (carrying energy, causing charged bodies to feel forces, pervading all of space), their states have a common cause (the behavior of charged bodies), and they stand in certain relations in virtue of the laws of nature.

In other words, classical physics fails to unify electricity and magnetism in the same sense as it unifies the forces governing the planets' and Moon's orbits, the tides, and the paths of bodies while falling to Earth. It is not merely the case that the Moon's orbit and a body's path while falling to Earth both involve bodies accelerating in response to forces in accordance with Newton's second law ($\boldsymbol{F} = m\boldsymbol{a}$) and that these forces have a common cause (the Earth's mass). The connection goes much deeper: The very same force, gravity, is responsible for both phenomena; classical physics reveals that *the Moon is just a falling body*, though from our vantage point, it certainly does not look that way. The only difference between the Moon and a stone falling to the Earth is that the stone quickly hits the Earth whereas the Moon keeps missing the Earth: as the Moon circles the Earth while falling towards it, the Earth's surface continually curves away beneath it. That a single force is responsible for the lunar and planetary orbits, the tides, and falling bodies constitutes a tremendous unification of what appear to be utterly different phenomena. In contrast, classical physics does not reveal electric forces and magnetic forces to be the same.[2]

2   How Relativity Unifies Electricity and Magnetism

A body that looks to be at rest, according to an observer keeping pace with it, may appear to be moving in a straight line at a constant speed from the vantage point of another observer. In the same way, a situation involving one combination of electric and magnetic fields, from one observer's perspective, will appear to another observer to involve a different combination of electric and magnetic fields. For example, consider two bodies some distance apart that carry the same electric charge and have been held at rest relative to a given observer (figure 7.1a). If we apply Maxwell's equations from this observer's vantage point, we find that each body repels the other electrically but exerts no magnetic force on the other. At each body's location, the electric field is non-zero but the magnetic field is zero. Now consider an observer from whose perspective the two bodies have been moving in parallel straight paths at the same constant speed perpendicular to the direction from one to the other (figure 7.1b). As moving charges, these bodies generate magnetic fields in addition to their electric fields. So

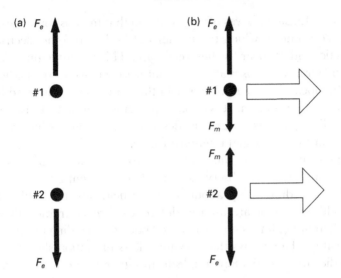

Figure 7.1
(a) From a vantage point relative to which the two bodies have been at rest, we see them as exerting equal and opposite electric forces on each other.
(b) But from a vantage point relative to which the two bodies have been moving to the right at equal speeds (white arrows show their velocities), the two bodies feel magnetic as well as electric forces.

applied from this viewpoint, Maxwell's equations say that the two bodies produce non-zero magnetic as well as electric fields at each other's locations, resulting in attractive magnetic and repulsive electric forces, the repulsion exceeding the attraction. The net repulsion from the second viewpoint is *unequal* to the repulsion measured from the first viewpoint.

In short, when Maxwell's equations are applied to the same situation as observed from different *frames of reference*, they may describe that situation as involving different combinations of electric and magnetic fields, giving rise to different net forces on the same bodies. If fields are interpreted as real entities, then the question arises: Applied from which frame of reference do Maxwell's equations describe the fields *as they really are*? Here is a roughly analogous question that has an obvious answer. Seen from above, a penny looks round. Seen from the side, the penny looks elliptical. The penny cannot really be both. So the penny really is the way it looks from (at most) one of these perspectives. Which perspective is that? (Or is it neither?) The answer, obviously, is that the penny really is the way it looks from directly above.

Prior to Einstein, it was often presumed that there is some particular reference frame in which the motions of bodies and the electric and magnetic fields appear as they really are. This was thought to be the reference frame that is really at rest (at least relative to the aether): an event's appearance to an observer in this frame is not distorted by the observer's own motion. It was often presumed, prior to Einstein, that Maxwell's equations accurately describe reality only when they are applied in this privileged reference frame.

Things are slightly different if fields are interpreted as mere calculational devices. Observations made in different reference frames are then not disagreeing about what the fields are really like. But if Maxwell's equations are used in different reference frames, they predict different net forces on the charged bodies depending on the frame. Classical mechanics says that Newton's laws of motion should be able to predict the observations made from either frame in our example (figure 7.1), and the frames should agree on the accelerations of all bodies – as I will discuss in chapter 8. Newton's second law says that a body's acceleration equals the net force on it divided by its mass. Since the two frames agree on the mass of every body, they must agree on the forces felt by the bodies. But Maxwell's equations predict different net forces in the two frames. So Maxwell's equations can be used to predict our observations only in one of the reference frames in which Newton's laws of motion can be used.

Einstein's special theory of relativity, on the other hand, is based on the assumption that Maxwell's equations and the other laws of nature can be applied (that is, accurately predict the observations that would be made) in *every* reference frame in which, *according to classical physics*, Newton's laws of motion are applicable. (Of course, relativity theory departs from classical physics in saying that Newton's laws do not hold exactly in any frame.) Moreover, relativity theory says that classical electromagnetic theory describes reality with equal accuracy from each of these viewpoints. So whatever is different from these different viewpoints is not real, but is at least partly a reflection of the viewpoint from which it is being measured.

In the example involving two bodies carrying equal charges, for instance, it is neither the case that there is really only an electric field nor the case that there is really an electric field and a magnetic field. There is only a single object, the electromagnetic field, and in different reference frames, it appears as different combinations of electric and magnetic fields.[3] Furthermore, for each of the two bodies, the two

Box 7.1
More on the example

Imagine that the experiment begins (at the firing of a pistol, say), then the bodies repel each other, and finally (at another firing of the pistol) the experiment is ended. Although the two frames disagree on the *rate of change* of a given body's momentum in the direction away from the other body, they agree on how much *total* momentum in that direction the body acquired during the experiment. That is because they disagree on how long the experiment lasted. As we shall see in chapter 8, the two frames disagree on the period of time between two events (such as the two pistol shots). That disagreement is just enough to compensate for their disagreement on the rate of momentum change, yielding the same net momentum change over the course of the experiment. Of course, relativity theory also says that the force in the first frame between (say) two electrons having long been at rest and separated by *r* in that frame equals the force in the second frame between two electrons having long been at rest and separated by *r* in that frame. That is because Coulomb's law, the law specifying an electron's charge, and the other laws are the same in both frames, according to relativity.

frames disagree on the force it feels, as well as on the resulting rate of change in its momentum. But in both frames, the same *relation* holds between the force and the rate of momentum change. Maxwell's equations and the other laws of nature hold in both frames. Neither is privileged in its depiction of reality (see box 7.1).

It is a familiar fact that a body's *velocity* in a given frame can be expressed as different combinations of components (figure 7.2). For the purposes of a certain calculation, one decomposition may be more convenient than another. But each of these combinations of components is an equally accurate representation of the speed and direction of the body's motion in that frame. Analogous remarks apply to the electric and magnetic components of the electromagnetic field, according to the theory of relativity. The same electromagnetic field can be accurately described in many frames of reference, and from these different viewpoints, it appears as different combinations of electric and magnetic components.

In a given frame, all decompositions of the same velocity will agree on certain facts. For example, suppose *v* is decomposed into three

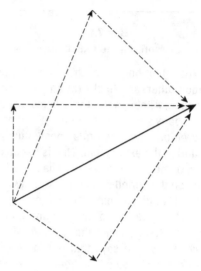

Figure 7.2 The same vector (solid arrow) can be decomposed into many different pairs of components (dashed arrows). Three pairs are shown.

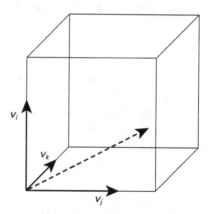

Figure 7.3 A velocity vector (dotted arrow) is the sum of three mutually perpendicular components (solid arrows).

components (v_i, v_j, and v_k) along axes at mutual right angles (figure 7.3). Then no matter what axes these are, the sum of the squares of these components (that is, $v_i^2 + v_j^2 + v_k^2$) is the same. Likewise, certain features are common to every decomposition of a given electromagnetic field into $\boldsymbol{E}$ and $\boldsymbol{B}$ components, where the different decompositions give the $\boldsymbol{E}$ and $\boldsymbol{B}$ combinations in different frames (in which, according to classical physics, Newton's laws of motion can be applied). For instance, in any of the reference frames, $E_iB_i + E_jB_j + E_kB_k$ is the same (where E_i, E_j, and E_k are $\boldsymbol{E}$'s components along any three mutually perpendicular axes, and B_i, B_j, and B_k are $\boldsymbol{B}$'s components along the same three axes). In other words, in any of these frames, $\boldsymbol{E}$'s magnitude times $\boldsymbol{B}$'s magnitude times the cosine of the angle between $\boldsymbol{E}$ and $\boldsymbol{B}$ is the same. This constant quantity is called the "dot product" of $\boldsymbol{E}$ and $\boldsymbol{B}$, symbolized $\boldsymbol{E} \cdot \boldsymbol{B}$. Another electromagnetic quantity that is the same in each of the frames is $E^2 - B^2$. These two quantities thus characterize the electromagnetic field *alone*; they in no way reflect the observer's perspective. Since they do not vary with the reference frame, they are called "invariants," as I shall explain further in chapter 8.

Bear in mind that Einstein was not responsible for naming his proposal "the theory of relativity." He would have preferred calling it "the theory of invariance," since it reveals what the genuine invariants are (Holton 1973: 362; Miller 1998: 163). Neither $\boldsymbol{E}$ alone nor $\boldsymbol{B}$ alone is an invariant. Only in combination do $\boldsymbol{E}$ and $\boldsymbol{B}$ correspond to some fact uncontaminated by our choice of a reference frame from which to describe the situation. If locality is satisfied, then the electromagnetic field is real, but the electric and magnetic fields are not; they do not exist independent of the reference frame from which we have chosen to describe things. (Since the Poynting vector $\boldsymbol{S} = (c/4\pi)\boldsymbol{E} \times \boldsymbol{B}$ is not an invariant, the electromagnetic energy flux is not an objective feature of the universe (Robinson 1995: 105–6). In chapter 8, we shall examine the consequences of this fact.) No matter what combination of $\boldsymbol{E}$ and $\boldsymbol{B}$ appear in a frame in which Newton's laws apply (according to classical physics), those fields obey Maxwell's equations.[4]

In short, magnetism turns out to be a relativistic phenomenon. A magnetic field is part of what an electric field (from one reference frame's perspective) looks like from another reference frame's vantage point. (As we saw in our example at the start of this section, a magnetic field disappears in a frame in which the charged body exerting it is at rest; in that frame, there is only the body's electric field. There is no frame in which that electric field disappears, since the body's charge is invariant. For more, see box 7.2.) Relativity theory explains how appearances from one frame's perspective relate to appearances from another viewpoint. That is why Maxwell's electromagnetic theory, in accurately covering electric and magnetic phenomena, is a piece of classical physics that was already relativistic before the advent of relativity theory. (Recall how relativistic we found equations (2.3)

Box 7.2
Magnetism as a relativistic phenomenon

Relativistic effects are more pronounced when large masses or high speeds are involved. There are no large masses or high speeds in magnetic phenomena. For example, a current-carrying wire is surrounded by a magnetic field. But the free electrons' "drift velocity" down the wire (which is superimposed on their random path in the wire, resulting from collisions with other particles) is only about 0.01–1 cm/sec. This hardly approaches c! However, although each free electron's magnetic contribution is tiny, there is roughly one free electron for each atom in a copper wire. This is a tremendous quantity: about 8.5×10^{28} free electrons per cubic meter. Separated by 1 m, the mobile electrons in two ordinary parallel wires would exert upon each other a total electric force of about 10^{16} newtons per meter of each wire! (Because conducting wires are net electrically neutral, there is no net electric force between them.) So we should not be surprised at relativistic effects arising in a reference frame relative to which *so much* charge is moving, albeit slowly.

and (2.4) to appear, though Heaviside derived them from Maxwell's equations.)

3 Einstein's Argument from Asymmetry

The unification of electricity and magnetism is not just a *consequence* of relativity theory. Einstein's original *motivation* for presuming that reality is accurately described from each of these reference frames is rooted in his hunch that electric and magnetic fields are not separate things. He suspected that the same single field is responsible for both the forces we call "electric" and the forces we call "magnetic":

> What led me directly to the Special Theory of Relativity was the conviction that the electromotive force induced in a body in motion in a magnetic field was nothing else but an electric field.   (Holton 1973: 285)

This "conviction" seems to have been based largely on an argument that Einstein gave in the opening lines of the 1905 paper in which he first set out the special theory of relativity.

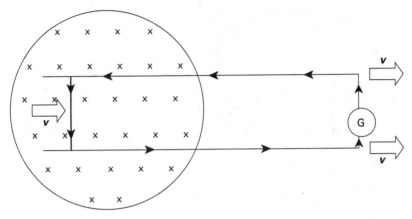

Figure 7.4 A solenoid (projecting above and below the page) produces a uniform
B field. (The magnetic lines of force, running into the page, are shown by ×s.) The
entire rectangle of conducting wires (including the galvanometer G and its wires) is
moving to the right at uniform velocity **v**. Electrons in the vertical rod in the field
feel a downward force, inducing them to flow counterclockwise around the circuit.
(An electron's **v** × **B** is upward, but since it is negatively charged, its (q/c)**v** × **B** is
downward.) The electron flow (shown by arrows) is measured by the galvanometer.
Though the top and bottom horizontal wires in the circuit are also moving, their
v × **B** is not directed along the length of the wire, and so the magnetic forces on
the electrons there are balanced by the forces keeping them in the wire. The wire
forming the right side of the circuit is located in a region where **B** = 0, and so no
magnetic force is felt by the electrons there.

The argument concerns cases like those we examined in chapter 2
in connection with the ontological status of lines of force. Consider a
vertical conducting rod attached to a return path (figure 7.4), where
the whole circuit is moving at a constant speed in a straight line through
a uniform unchanging magnetic field, as exists inside a solenoid through
the coils of which a constant current flows (Miller 1998: 138–42;
Halliday et al. 1992: II, 793–5). Since **B** is unchanging, **B**′ = 0, and so
no **E** is induced according to Faraday's law of electromagnetic induction (**curl E** = −(1/c) **B**′). Since the free electrons in the rod are
moving, their **v** is non-zero, and so they feel a magnetic force down-
ward in figure 7.4 in accordance with $\boldsymbol{F}_m$ = (q/c)**v** × **B**. Consequently,
the electrons flow around the circuit counterclockwise.

Compare this case to a situation where the wire and magnetic field
have exactly the same relative motion as before, but now the field (that
is to say, the solenoid) is moving while the wire is still (figure 7.5). Again,

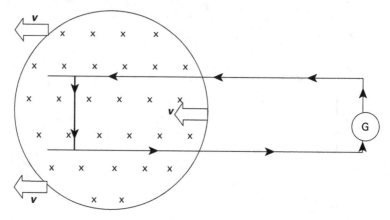

Figure 7.5   The same as the preceding figure, except now the conducting wires are stationary and instead the solenoid (with its magnetic field) is moving to the left at uniform velocity v.

we apply Maxwell's equations in the reference frame that is at rest (at least relative to the aether). Since the wire's $v = 0$, the magnetic force, $F_m = (q/c)v \times B$, on the electrons in the wire is zero, in contrast to the earlier case. However, B' is non-zero at any location at the edge of the solenoid's field, since these locations are just receiving a non-zero B as the solenoid reaches them or are just returning to a zero B as the solenoid leaves them behind. So at these locations – which coincide with the solenoid's own windings – there is a non-zero electric field, according to Faraday's law. As a result, the top half of the solenoid becomes electrically negative, the bottom half electrically positive, and the resultant electric field makes the rod's free electrons move downward, initiating a counterclockwise flow – just as in the earlier case. Indeed, when v is small compared to c, E in this case is equal to $(1/c)v \times B$ in the first case, and so the force on the rod's electrons is the same in both cases. Yet the force is magnetic in the first case and electric in the second; a magnetic field but no electric field is present in the first case, whereas both non-zero electric and magnetic fields exist in the second case.

   Let us look at another pair of cases that has the same interesting features as the pair we just examined (Becker 1964: 380–2; Miller 1998: 152–4). Consider an iron bar magnet with long ends at the poles (figure 7.6). Because the magnet's north pole (for instance) is so long, there is a large region above the pole in which the magnetic lines of force (emerging from the magnet) remain straight and parallel.

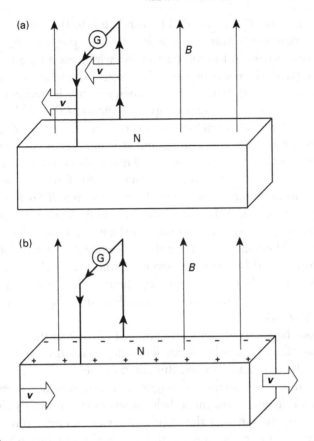

Figure 7.6
(a)   In sliding contact with the north pole of a squat magnet, a loop of wire
(including a galvanometer) moves to the left at uniform velocity **v**. Some of the
magnet's lines of force are shown emanating from its north pole. Arrows along the
wire loop depict the electron flow that is induced around the circuit. (One leg of
the circuit lies along the magnet's upper surface.)
(b)   The same as the preceding figure, except now the galvanometer and its wires
are stationary and the magnet underneath them moves to the right at uniform
velocity **v**. Charges are induced on the magnet's upper surface.

Suppose we use sliding contacts to hook the ends of a conducting wire
to opposite sides of the magnet's north pole. Let us add a galvanometer
to measure any current induced to flow through the wire. Now consider two cases. First, suppose that the magnet is stationary while the
galvanometer circuit slides to the left at a modest speed v (figure 7.6a).
Since the magnet is unmoving, **B** is nowhere changing, and so (by

Faraday's law) no E is induced. The wire's v is to the left and B at its location is pointing upward, so $v \times B$ is into the page. By $F_m = (q/c)v \times B$, a free electron in the top part of the wire feels a magnetic force out of the page (since q is negative for electrons) and a current flows. Now suppose instead that the galvanometer circuit is stationary while the magnet slides to the right at speed v (figure 7.6b). Although the magnet is moving, B is unchanging everywhere near the wire's location (since the magnet is very long and B is uniform in a large region above the pole). So Faraday's law thus far reveals no induced E. Since the wire's $v = 0$, its free electrons feel no magnetic force. However, the free electrons *in the magnet* have v to the right, so $v \times B$ for them is out of the page. Consequently, they feel a magnetic force into the page. Hence, the magnet polarizes: its inner surface (where the electrons are accumulating) becomes negatively charged while its outer surface (with an electron deficit) becomes positively charged. The electrons in the wire thus feel an electric force causing them to flow from negative to positive – a current of the same magnitude and in the same direction as in the first case.

As in our first pair of cases, the induced current in these two cases (known as "linear unipolar induction") depends only on the *relative* motion of the magnet and wire. But whether the wire's electrons were caused to move by electric or magnetic forces, whether there is an electric field or only a magnetic field, whether the magnet's electrons are displaced, and whether the "seat of the electromotive force" is the galvanometer wire or the magnet (that is, whether the $(q/c)v \times B$ force is felt by free electrons in the wire or the magnet) – all of these facts depend, in classical physics, on whether the magnet or the wire is *really* moving. They depend, in other words, on facts about absolute motion over and above relative motion.

What did Einstein find so puzzling or unsatisfactory about this picture? Here are the majestic opening words of Einstein's first paper on relativity, from 1905:

> It is well known that Maxwell's electrodynamics – as usually understood at present – when applied to moving bodies, leads to asymmetries that do not seem to attach to the phenomena. Let us recall, for example, the electrodynamic interaction between a magnet and a conductor. The observable phenomenon depends here only on the relative motion of conductor and magnet, while according to the customary conception

the two cases, in which, respectively, either the one or the other of the two bodies is the one in motion, are to be strictly differentiated from each other. For if the magnet is in motion and the conductor is at rest, there arises in the surroundings of the magnet an electric field endowed with a certain energy value [namely, an energy density of $E^2/8\pi$, as we saw in chapter 5] that produces a current in the places where parts of the conductor are located. But if the magnet is at rest and the conductor is in motion, no electric field arises in the surroundings of the magnet, while in the conductor an electromotive force will arise, to which in itself there does not correspond any energy [because there is no electric field], but which, provided that the relative motion in the two cases considered is the same, gives rise to electrical currents that have the same magnitude and the same course as those produced by the electric forces in the first-mentioned case.

Examples of a similar kind, and the failure of attempts [such as the famous one by Michelson and Morley, which I shall mention again in chapter 8] to detect a motion of the earth relative to the "light medium" [the aether], lead to the conjecture that not only in mechanics, but in electrodynamics as well, the phenomena do not have any properties corresponding to the concept of absolute rest, but that in all coordinate systems in which the mechanical equations are valid, also the same electrodynamic and optical laws are valid. . . . We shall raise this conjecture (whose content will be called "the principle of relativity" hereafter) to the status of a postulate [which I shall explain further in chapter 8]. (1989: 140)

Einstein's eventual conclusion is that in each pair we have just examined, the two cases are really not at all different. There is no fact about whether the magnet is moving and the wires are at rest, or vice versa. There is no fact about whether both electric and magnetic fields are present, or only a magnetic field. There is no fact about whether the forces causing the current to flow are electric or magnetic. Rather, the only facts are the *relative* motions of the magnet and wire, as well as the existence of an electromagnetic field characterized by the invariant quantities I mentioned earlier involving both **E** and **B**. Seen from different reference frames, the electromagnetic field appears as different combinations of **E** and **B**.

But how was Einstein led to this view by examples like the ones we have looked at? The 1905 paper merely complains of "asymmetries that do not seem to attach to the phenomena." Einstein later explained his motivation in these words:

In the construction of special relativity theory, the following ... played for me a leading role.

According to Faraday, during the relative motion of a magnet with respect to a conducting circuit, an electric current is induced in the latter. It is all the same whether the magnet is moved or the conductor; only the relative motion counts. ... However, the theoretical interpretation of the phenomenon in these two cases is quite different. ...

The thought that one is dealing here with two fundamentally different cases was for me unbearable. The difference between these two cases could not be a real difference but rather, in my conviction, only a difference in the choice of the reference point. Judged from the magnet, there were certainly no electric fields, [whereas] judged from the conducting circuit there certainly was one. The existence of an electric field was therefore a relative one, depending on the state of motion of the coordinate system being used, and a kind of objective reality could be granted only to the *electric and magnetic* field together, quite apart from the state of relative motion of the observer or the coordinate system. The phenomenon of the electromagnetic induction forced me to postulate the (special) relativity principle.   (Holton 1973: 363–4; see Stachel 1987: 262–3)

How should we understand Einstein's finding it "unbearable" to give different interpretations to two cases involving the same relative motion and the same resulting current?

One possibility is that Einstein's remarks regarding "asymmetries that do not seem to attach to the phenomena" do not express a *reason* for being dissatisfied with electromagnetic theory as it was then "usually understood." That is, Einstein's remarks do not supply an *argument* for the proposal that Einstein goes on to offer, but merely reflect Einstein's *faith* that the universe is organized in a certain fashion that he finds aesthetically pleasing (Holton 1973: 289; Miller 1982: 390). As recalled by Hans Reichenbach (German-American philosopher of science, 1891–1953, who was a student in the first relativity course Einstein taught at the University of Berlin – only 5 students were enrolled!):

When I, on a certain occasion, asked Professor Einstein how he found his theory of relativity, he answered that he found it because he was so strongly convinced of the harmony of the universe. No doubt his theory supplies a most successful demonstration of the usefulness of such a conviction. But ... [t]he philosopher of science is not much interested in the thought processes which lead to scientific discoveries; he looks for a logical analysis of the completed theory, including the relationships establishing its validity. That is, he is not interested in the context of discovery [Reichenbach's term for the various factors, personal or sociological, that originally caused the scientist to think up the theory] but in the context of justification [Reichenbach's term for the arguments by which the theory is supported or disconfirmed by evidence]. (1949: 292)

In other words, Einstein's "personal faith" (Reichenbach 1949: 293) in the universe's harmony, though having no logical basis, was an aspect of Einstein's character that, as luck would have it, led him toward the truth. That relativity theory would eliminate an "asymmetry" from classical electromagnetic theory did not join various experiments regarding the impossibility of measuring a body's absolute velocity (that is, regarding light's having the same speed in different frames) as *evidence* for relativity theory.

That these "asymmetries" supplied no *reason* for being dissatisfied with the usual interpretation of electromagnetic theory has sometimes been used to explain why no other physicist (save one, whom I'll mention) had voiced similar dissatisfaction (Holton 1973: 289; Darrigol 1996: 256). Ample time had elapsed for such dissatisfaction to be felt: The fact that the resulting current in the cases we have examined depends only on the *relative* motions of the magnet and conductor had been well known since Faraday's original experiments on electromagnetic induction in 1831. Yet even Heaviside did not make the "asymmetry" argument against the standard interpretation of electric and magnetic fields. He examined a pair of cases involving the same relative motions and different absolute motions. He noted that there would be the same induced effect in both cases and then blandly commented that "[o]therwise, however, there is a great difference in the two experiments" (1925: I, 445; see Darrigol 1996: 262). He did not seem to find the existence of these differences problematic.

Nevertheless, despite the subjective terms in which Einstein frequently puts the point (as when he says that the conventional interpretation "was for me unbearable"), it is hard to avoid concluding that Einstein regarded these asymmetries as supplying a powerful *argument* against what was then the conventional wisdom.[5] A mere psychological impetus toward relativity theory, a guiding inspiration having no force as an argument in favor of the theory's truth, would not have belonged in the first scientific paper advancing the theory (much less as its opening lines). As an undergraduate, I heard that a famous astrophysicist at our school had been inspired to develop a certain theory of the universe's four-dimensional shape by an experience he once had with an ice-cream cone in the ice-cream parlor near our campus. This episode, however much a factor in setting the astrophysicist's mind working in a particular (and ultimately fruitful) direction, would be utterly out of place in a scientific paper defending the cosmological theory. In Reichenbach's terminology, this experience belongs to the "context of discovery" rather than to the "context of justification." Einstein apparently takes himself to be *arguing* against classical electromagnetic theory (as it was then usually interpreted) on the grounds that "an asymmetry of the theoretical structure, to which there is no corresponding asymmetry in the system of empirical facts, is intolerable to the theorist" (1934: 103).

How is this argument supposed to go? Here is one interpretation of it (roughly along the lines of Norton 1991: 135–6). Any justification for believing that uniform absolute motions (over and above uniform relative motions) make any difference to electromagnetic phenomena must ultimately come from observations. But observations reveal no differences (no "asymmetry") between cases involving the same uniform relative motions but allegedly different uniform absolute motions.[6] The electromagnetic phenomena we observe (such as induced currents) depend only on the *relative* uniform motions of conductors, magnets, and so forth. Thus, our observations do not supply us with a good reason for believing that differences in uniform *absolute* motion make any difference to electromagnetic results. Insofar as classical electromagnetic theory portrays certain electromagnetic results (such as the kind of force making a current flow) as depending upon uniform absolute motions over and above uniform relative motions, that theory is unjustified. To be justified by our observations, our theory must avoid hypothesizing any differences between the two cases in one of our pairs.

Einstein's argument is plainly concerned with how far our observations go in justifying classical electromagnetic theory as it was then commonly interpreted.

> [I]t disturbed me that electro-dynamics should pick out *one* state of motion [as at rest relative to the aether] in preference to others, without any experimental justification for this preferential treatment. (1934: 113)

Einstein also wrote:

The type of critical reasoning which was required for the discovery of this central point was decisively furthered, in my case, especially by the reading of David Hume's and Ernst Mach's philosophical writings.   (1949: 53)

Hume and Mach (Moravian physicist and philosopher, 1838–1916) were famous for taking a ruthlessly critical attitude toward various commonsense beliefs, scrupulously investigating whether they are really justified by our observations. Einstein apparently took himself to be following their example in holding that even our commonsense ideas about space and time (which may at first appear beyond question) can be overturned on the basis of observational evidence.

[Mach] has convincingly advocated the point of view that [concepts in physics], even the most fundamental, receive their justification only from experience, that they are in no way *logically* necessary.   (Hirosige 1976: 60; see Einstein 1953: 2)

However, that a theory can be justified only by observations does *not* mean that any "asymmetry" in our theoretical account of the two cases in one of our pairs can be justified only by some difference in the observed features of these cases. This would require an excessively direct connection between observations and what they can justify. That there are no observable differences between cases involving the same uniform relative motions, but allegedly different uniform absolute motions, does not entail that observations fail to *justify* belief in a theory that describes those cases differently. Observations can justify belief in a theory possessing certain characteristics even if the observations themselves lack those characteristics. For instance, although the Moon's orbital behavior is very different from the trajectory of a ball falling to the ground, observations support a theory that describes the Moon as just another falling body. If observations can justify a theory in holding that there are hidden unities in phenomena that appear distinct, then observations can justify a theory in holding that there are hidden differences among apparently identical cases.

Still, that a theory gives different treatments to certain observationally identical cases would appear at least to put that theory at a competitive disadvantage – to make it harder to justify. If a given difference hypothesized by the theory is hidden from direct observation, then some less direct means of confirming its existence must be found, or else belief in the theory will not be justified. According to Einstein, there need not be a direct connection between the various elements of a theory and the various observations that together would suffice to justify the theory. For a theory to be justified,

> it is only necessary that enough propositions of the [theory] be firmly enough connected with sensory experiences and that the [theory], in view of its task of ordering and surveying sense-experience, should show as much unity and parsimony as possible.   (1944: 289)

Perhaps, then, the point of Einstein's "asymmetry" argument is that classical electromagnetic theory does not "show as much unity and parsimony as possible" because it describes the two cases in one of our pairs as differing in the hidden absolute motions of the system's components. A distinction between the two cases is unjustified because it amounts to excess baggage – considering the indifference to absolute motion displayed by the observations being explained.

Perhaps this thought also motivated the only researcher before Einstein (to my knowledge) who was bothered by the "asymmetry" Einstein mentions: the telegraphic engineer S. Tolver Preston, whose insightful 1885 critique of Faraday's lines of force we encountered in chapter 2. In the same paper, Preston considers what happens when a cylindrical magnet rotates around its long axis, with galvanometer wires in sliding contact with its pole and its middle (figure 7.7). As Faraday (1839: 65; 1935: I, 301) had reported in 1832, a current is induced in the galvanometer wires (a phenomenon later called "[rotatory] unipolar induction"). On Faraday's view, the lines of force do not rotate with the magnet; the current is induced by the magnet cutting its own lines of force. On Preston's view, the lines of force are rigidly attached to the magnet; the current is induced by the stationary galvanometer wires being cut by the rotating lines of force. Now consider what happens when the magnet is at rest and the galvanometer wires revolve so that the relative motion of the magnet and wires is the same as in the first case. The same current is induced. On Faraday's view, the current this time is caused by the galvanometer wires cutting the lines of force, whereas in the first case, it was caused by the rotating magnet cutting its own (stationary) lines. Preston sees his theory (according to which the lines are attached to the magnet) as *superior* in attributing the current in each case to the same cause – the galvanometer wires cutting the lines:

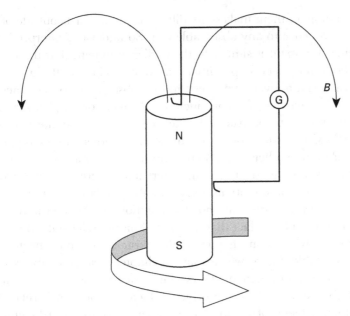

Figure 7.7   Galvanometer wires in sliding contact with a rotating cylindrical magnet. Two magnetic field lines are shown emanating from the magnet's north pole.

> This explanation has the . . . advantage [this sure sounds like it's meant to be an argument!] of bringing the current formed, whether by the rotation of the magnet on its axis, or by the (inverse) revolution of the wire about the magnet, *under the same cause* (and not under different causes).   (1885: 135)

This may be just the sort of "unity and parsimony" to which Einstein referred. Of course, Preston's picture of lines of force as real entities did not displace Faraday's picture (which grew into Maxwell's field theory) or cope with later experiments (as we saw in chapter 2). Yet the aspect of Faraday's picture that originally disturbed Preston grew into the asymmetry in field theory that later disturbed Einstein.

This argument from what Redhead (1975: 86–7) calls "surplus structure" might be compared to the argument we saw in chapter 2 against the reality of the electric potential: that its absolute value would be a "dangler," a loose end, making no difference to any other facts. The wire's absolute motion (at least relative to the aether) is not a dangler according to classical electromagnetic theory, since it makes a difference to whether certain *other* facts hold – for example, to whether a certain force is electric or magnetic. But this *group* of facts forms a dangler: none of them makes any difference to any facts outside of the group, and hence to any observable fact. So under this interpretation, Einstein's argument is similar to the earlier argument that there is no fact about the electric potential's absolute value: both arguments emphasize that a given hypothesis has no observable consequences.

Does the fact that a theory includes danglers or surplus structure make it unlikely to be true? On the one hand, if absolute values of potential existed, why would they make no difference to anything observable? On the other hand, there is no reason why absolute values of potential would have to make an observable difference. One way to explain why absolute motions (over and above relative motions) make no difference is that there are no absolute motions. An alternative view is that it is just a brute fact that the laws of nature refer only to relative motions. As we saw in chapter 4, the existence of some brute facts seems inevitable. But as we also saw in chapter 4, some relations seem more plausibly brute than others – in other words, less fine-tuned. Recall, for example, how the orbital velocity of the Sun is related to the orbital velocity of a superior planet in Ptolemy's Earth-centered model of the solar system. This relation just happens to result in the superior planet always reaching its nearest point to Earth (its "perigee") exactly when it arrives on the opposite side of the Earth from the Sun ("opposition"). The exquisite alignment of the model's free parameters (the orbital velocities) that was needed to yield the observed correlation between oppositions and perigees seemed to Kepler to constitute suspicious fine-tuning – something probably having a deeper explanation, not the sort of fact that is likely to be brute. The correlation between the superior planets' oppositions and perigees was indeed explained, without any such fine-tuning, on Copernicus's Sun-centered model of the solar system.

Likewise, with regard to electromagnetic induction, there is something suspicious about a theory according to which there are two entirely different things going on, depending on whether the magnet or the wire is really moving, but nevertheless the *very same current* is induced by both processes so long as the bodies' relative motions are the same. If absolute motion made a difference to the hidden processes inducing the current, then why wouldn't it make a difference to the amount or direction of the current induced? It might make no difference just as a matter of brute fact, but considering that the hidden processes inducing the current are quite different in the two cases, it would have to be judged a remarkable coincidence that the current induced when the conductor is moving and the magnet is stationary turns out to be exactly equal to the current induced when the conductor is still and the magnet is moving – as if the laws of nature were participants in a giant conspiracy to hide the absolute motions from us.

The unlikelihood of such a conspiracy (that is, of two independent processes leading to identical results) is also behind Einstein's argument against the "Lorentz–FitzGerald contraction hypothesis." According to that hypothesis, the reason why light takes the same amount of time to travel the length of a body, whether the body is at rest or in absolute uniform motion in the direction that the light is traveling, is because the body's motion contracts it by just the right amount to compensate for the body's running away from the light. Einstein wrote:

> This hypothesis formally suffices for the facts of the situation, but in spite of that the mind remains dissatisfied. Is nature really supposed to have placed us in an aether gale [because the Earth is moving relative to the aether], and on the other hand exactly so arranged the laws of nature that we can notice nothing of this gale [because bodies on Earth are contracted by the right amount to compensate for their motion through the aether]?   (Stachel 1982: 52)

This would be an unlikely coincidence in the laws, a remarkable case of "fine-tuning" between two independent processes.[7] It cries out to be eliminated – as Einstein does by eliminating the differences in absolute motions between cases involving the same relative uniform motions.

On this interpretation, the differences in absolute motion between the two cases in one of our pairs are not merely surplus structure – more than the minimum necessary to accommodate our observations. That the two electromagnetic processes resulting from these different absolute motions happen to generate exactly the same currents represents suspicious fine-tuning, like the relation between the orbital velocities of the Sun and superior planets in the ancient Earth-centered solar-system model. This appears to be the fundamental idea behind Einstein's asymmetry argument.

4   The Interdependence of Philosophy and Physics

The asymmetry Einstein presented was a good reason for taking seriously a theory that unified electric and magnetic fields. This does not mean that all by itself, the asymmetry argument was sufficient to justify believing in Einstein's theory; despite Einstein's argument, a scientist at the time might still have been justified in attributing these asymmetries to a coincidental feature of the natural laws. Not all arguments are decisive – so strong that to resist their conclusions must amount to prejudice, bias, closed-mindedness, or some other sort of failure to weigh the available evidence fully and impartially. On the other hand, in the face of Einstein's argument, a scientist who disagreed would have to have some *reasons* for resisting Einstein's proposal. It could not merely be brushed aside as obviously unworthy of being taken seriously.

Thus, Einstein's finding the asymmetry "unbearable" was in fact not a mere aesthetic preference. It belongs to relativity's "context of justification," not solely to its "context of discovery." That Einstein and Preston alone saw the asymmetry as something to worry about does not suggest that their dissatisfaction with the asymmetry arose from a rare, mystical faith that a certain kind of harmony prevails. Rather, that only Einstein and Preston were troubled by the asymmetry shows that tremendous sensitivity and insight were required to recognize that a certain well-known fact (that the magnitude of the induced current depends only on the relative uniform motions of conductors and magnets), although accurately predicted by a popular and otherwise well-supported theory, actually supplies a good reason to suspect that the theory (under its usual interpretation) is mistaken. As Einstein wrote in a different context:[8]

> It is really rather strange that human beings are normally deaf to the strongest arguments while they are always inclined to overestimate measuring accuracies. (Born 1971: 192)

Einstein's asymmetry argument, however its details should be understood, obviously does not question the *accuracy* of classical electromagnetism's predictions regarding induced currents and other phenomena we have observed. Rather, Einstein's aim is to criticize a certain *interpretation* of classical electromagnetism – a certain account of what reality must be like considering classical electromagnetism's accuracy regarding our observations. Obviously, Einstein's interpretive project eventually proved to be immensely fruitful for physics, ultimately leading to a new mechanics with greater accuracy regarding our observations than Newtonian mechanics possesses. The interpretation of physical theories is thus capable of leading to new scientific theories, making novel predictions regarding our observations. Of course, this moral was also implicit throughout chapters 2 and 5, where we witnessed Faraday, Maxwell, Heaviside, Lodge, and Poynting arriving at electromagnetic field theory partly as a result of their struggles with interpretive, philosophical questions about how charged bodies manage to act on one another despite being separated in space and time.

The interpretation of physical theories is not an activity pursued in parallel to but outside of physics, as a kind of extracurricular activity for working scientists. Rather, the philosophical interpretation of scientific theories should be understood as integral to at least some scientific work.

Discussion Questions

You might think about . . .

1   According to Einstein's recollections, relativity theory

> resulted from a paradox upon which I had already hit at the age of sixteen: If I pursue a beam of light with the velocity c . . . , I should observe such a beam of light as a spatially oscillatory electromagnetic field at rest. However, there seems to be no such thing, whether on the basis of experience or according to Maxwell's equations.   (1949: 53)

But what is the least bit paradoxical here? That we have never observed anything like a spatially periodic electromagnetic field at rest is to be expected if we have never traveled at speeds (relative to the aether) near c. And if Maxwell's equations accurately describe reality only when they are applied in a reference frame that is at rest relative to the aether, then we should expect things to appear to a moving observer to violate Maxwell's equations (Darrigol 1996: 289–90). Darrigol notes (p. 264) that "Heaviside [1922: I, 44] performed a similar thought experiment [to Einstein's], but detected no paradox." Is Einstein's famous argument from "riding on a beam of light" flawed?

2   It is sometimes said that the tides, falling bodies, and planetary motions are governed by the same force. What does this mean *when no*

mechanism under which that force operates has been specified? Does it mean, for instance, that all of these phenomena are governed by the same law of nature (namely, the gravitational force law)? But when do you have just one law as opposed to many? Feynman writes (sarcastically):

> Let us show you something interesting that we have recently discovered: *All of the laws of physics can be contained in one equation.* That equation is
>
> $$U = 0.$$
>
> What a simple equation! Of course, it is necessary to know what the symbol means. U is a physical quantity which we will call the "unworldliness" of the situation. And we have a formula for it. Here is how you calculate the unworldliness. You take all of the known physical laws and write them in a special form. For example, suppose you take the law of mechanics, $F = ma$, and rewrite it as $F - ma = 0$. Then you can call $(F - ma)$ – which should, of course, be zero – the "mismatch," of mechanics. Next you take the *square* of this mismatch and call it U_1, which can be called the "unworldliness of mechanical effects." [You continue doing this for all other physical laws, producing U_2, U_3, and so on, one for each law.] Finally, you call the *total* unworldliness U of the world the sum of the various unworldlinesses U_i from all of the subphenomena that are involved. . . . Then the great "law of nature" is $U = 0$. This "law" means, of course, that the sum of the squares of all the individual mismatches is zero, and the only way the sum of a lot of squares can be zero is for each one of the terms to be zero.   (Feynman et al. 1963: II, pp.25-10 to 25-11)

Why, precisely, does this ruse not succeed in unifying everything under a single law? Is there any difference between the bogus "unity" achieved by this ruse and the alleged unity achieved by Newton's theory of gravity *when no mechanism for gravity has been specified*? If so, what is the difference?

3   An experiment was performed to test the policy of giving unemployment payments to a felon for the first 6 months after her release from prison or until she has found a job, whichever comes first. One randomly chosen group of newly released felons was given unemployment payments. The other group was not; instead, the newly released felons were employed. The rates at which the two groups were re-arrested was found to be the same. How should this result be interpreted?

One possibility is that being unemployed increases the likelihood that a former felon will return to a life of crime, whereas receiving unemployment payments exactly counterbalances that effect, so that the net result of these two opposing influences is that the likelihood of rearrest among the unemployed former felons receiving unemployment payments is the same as the likelihood of rearrest among employed former felons. Another possibility is that neither unemployment nor unemployment payments have any effect on the likelihood of a released felon's committing another crime. (This hypothesis is defended by Glymour et al. 1987: 26.) Can a good argument be given for the second hypothesis along the same lines as Einstein's asymmetry argument? Or would it be less mysterious in this case why two apparently separate mechanisms (employment vs. unemployment with unemployment payments) give exactly the same results?

Notes

1   Much has been written about Maxwell's diverse attitudes toward the various physical models he employs. See Siegel (1985).
2   There is a tremendous literature on what unification is – or, perhaps more accurately, on different senses or aspects of "unification" in science. See Friedman (1974), Kitcher (1989), Maudlin (1996), Morrison (2000).
3   Contrast Feynman et al. (1963: II, pp.1-10 and 13-10), who say that lines of magnetic force are not real because they would have to vanish in certain frames. This frame-dependence is a good reason for denying $\boldsymbol{B}$'s reality. But the lines of force were already shown to be unreal without appealing to relativity theory – back in chapter 2.
4   Maxwell's equations do not hold in a rotating frame. To apply them there, you need to introduce fictitious charges and currents (determined by Schiff's equations, derived from the general theory of relativity), just as Newton's laws apply in a rotating frame only when fictitious forces are introduced.
5   Einstein says that there are two components to judging the truth of a theory, the first concerned with agreement with observation, the second "with what may briefly but vaguely be characterized as the 'naturalness' or 'logical simplicity' of the [theory]" (1949: 23).
6   The same cannot be said for accelerated motions (in particular, for rotations) as we saw in the case of Faraday's disk. See note 4.
7   What is the basis of our belief that this sort of fine-tuning is unlikely? Einstein doesn't say (and in chapter 4, we saw how difficult this question can be). One possibility is that our belief is based on what we have

learned about the way the universe happens to be. On this interpretation, Einstein's argument is not based on some supposed logical principle concerning observables, unobservables, and theory testing, such as the principle that belief in unobservable differences without observable differences is always unjustified. (Compare Norton 1991.) Rather, Einstein's argument turns on what we have discovered regarding the rarity of certain sorts of brute coincidences (whether among observables or unobservables). On the other hand, Laurence BonJour has suggested that all inductive reasoning is based on the supposition that a certain sort of coincidence is unlikely: that all Fs that have heretofore been checked have turned out to be G although not all actual Fs are G. On this view, the principle behind Einstein's argument lies behind every inductive argument, and so we cannot arrive at it by reasoning inductively from our observations.

8   Einstein is contending that there would have been powerful arguments favoring the general theory of relativity even if no observations had yet been made of the extreme cases in which Einstein's theory predicts different observations than its rivals – indeed, even if initial observations had disagreed with Einstein's predictions.

8

Relativity, Energy, Mass, and the Reality of Fields

Einstein's theory of relativity contains fundamental insights regarding many of the issues that we have been investigating in connection with spatiotemporal locality.[1] In the previous chapter, we saw how relativity unites electricity and magnetism. We shall now examine other cases in which relativity unifies concepts that in classical physics remain utterly separate: time and space, and especially energy, momentum, and mass. We shall see why in relativity theory, a body's mass cannot be interpreted as the amount of matter that the body is made of. We shall also see why it is incorrect to interpret relativity theory as saying that mass is sometimes *converted* into energy – as saying, for instance, that in a nuclear fission reaction, some quantity of mass turns into energy, resulting in a "mass defect" that violates the classical law of mass conservation.

In this connection, we will have to figure out the proper way to interpret "mass-energy equivalence" and Einstein's famous formula $\mathcal{E} = mc^2$. This formula has often been interpreted as expressing the fact that mass is a form of energy or that mass and energy are really the same thing ("mass-energy"). This common interpretation is, at best, highly misleading. The "conversion" of mass into energy is not a real process like the metamorphosis of a caterpillar into a butterfly. Having sorted all of this out, we will be well positioned to shed some light on questions about the reality of energy, its distribution in an electromagnetic field, and the Poynting vector's correctness in depicting energy flow – questions that we could not fully resolve in chapter 5. These results will, in turn, help us to understand whether the electromagnetic field is a real thing and, hence, whether electromagnetic interactions obey spatiotemporal locality.

1　Classical Physics and the "Relativity of Motion"

A body's path assumes different forms in different "frames of reference." For example, imagine two people throwing a ball back and forth as they sit, side by side, in the rear seat of a car. From the vantage point of a reference frame attached to the car, the ball retraces the same straight line, over and over. However, imagine another reference frame relative to which the car is moving at a constant speed to the right. In that frame, the ball does not retrace the same line, since the ball is moving not only across the car's width, but also to the right along with the car (figure 8.1).

In setting up a coordinate system in terms of which to describe a body's motion, does it matter where we put the origin (the point labeled "0,0,0") at a given time? Of course, one location might be more *convenient* than another to take as the origin. But we cannot make an *error of fact* in selecting one point rather than another as our origin. Likewise, it is arbitrary which moment we choose to identify as $t = 0$ – the "origin" for measurements of time.

In contrast, there *is* a fact about my distance now from you. According to classical physics, this quantity is the *same* in every reference frame in terms of which the universe can be accurately described (in that all of the universe's features can be correctly represented in a coordinate system laid down in that frame). In other words, the separation between two bodies is an *invariant* quantity in classical physics. In contrast, there is no fact about which of us is nearer now to the origin except *relative to* a given reference frame (and, furthermore, relative to a given coordinate system in that frame). In other words, my distance from the origin is a *frame-dependent* quantity. A real quantity must be invariant. The value that any frame-dependent quantity takes

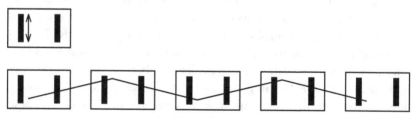

Figure 8.1　In the upper figure, the line with arrows depicts the ball's path as seen from directly above the car. The lower figure depicts the car at five moments while it is traveling to the right. The line depicts the ball's path in this reference frame.

on in a given reference frame reflects not just reality, but also our decision to use that particular reference frame (or perhaps even to use a certain coordinate system in that frame). An invariant quantity, on the other hand, characterizes reality alone, uncontaminated by anything contributed by us in the course of describing reality. If two frames from which the universe can be accurately described disagree on a certain matter, then that matter cannot be an objective fact.

We can accurately describe a "snapshot" of reality – the universe *at a single moment* – no matter where we take the origin to be. But matters are more complicated when we try to describe the universe *over time*, as when we try to describe a body's motion. Suppose that from frame A's standpoint, each point at rest in frame B moves in a continuous path, but not uniformly: it speeds up, slows down, and changes direction. Hence, these two frames disagree on whether a given body is accelerating. For example, a body at rest according to B is accelerating according to A. Are both of these systems accurate, or is there an objective fact about whether a given body is accelerating?

The answer is given by the laws of nature. Newton's laws of motion, the foundation of classical physics, distinguish acceleration from constant velocity (that is, rest or motion in a straight line at a constant speed). Newton's first law, for instance, says that a body undergoes no acceleration if it feels no forces. This law is meaningless if there is no objective fact about whether a body is accelerating.

Acceleration is therefore (according to classical physics) an invariant quantity: every reference frame from which the universe can be accurately described must agree on whether or not a given body is accelerating. Since frames A and B disagree on whether a given body is accelerating, either A or B does not permit the universe to be described accurately. A frame that portrays as accelerated all and only the bodies that really are accelerated is called *inertial*. That is because in those frames, Newton's first law (the principle of inertia) holds: a body acted upon by no forces undergoes no acceleration. Classical physics says that in all and only the inertial frames, the motions of bodies are governed by Newton's laws.

For example, in a reference frame attached to the Earth, some bodies undergo acceleration despite feeling no forces that could account for this acceleration through Newton's second law ($F = ma$, where F is the net force on a body, m is its mass, and a is its acceleration). For example, water going down a bathtub drain in the northern hemisphere rotates counterclockwise, but nothing is pushing on it to make it turn. Since Newton's laws fail to govern the motions of bodies in a frame attached to the Earth, the Earth must be accelerating, as Copernicus believed.

There is an objective fact regarding whether a given body is accelerating, and its acceleration is defined as its velocity's rate of change. But *it does not logically follow* that there is an objective fact regarding a body's velocity (just as a field vector corresponds to potential differences, but the field's reality does not require the potential's reality). Newton's laws distinguish acceleration from non-acceleration, but they do not distinguish unaccelerated motion from rest.

Newton's physics, then, leaves room for two rival views regarding the objectivity of a body's velocity. Newton's own view was that not all inertial frames can be used to describe the universe accurately. There is an objective fact regarding which inertial frame portrays as moving all and only the bodies that really are moving. This is the inertial frame that is *really* at rest. A body's velocity in that frame is its *absolute* velocity: its velocity relative to the points of space, which are really at rest. These points are real entities with continuing identities over time: there is a fact about whether the point where the Earth's center is now located is the selfsame point as the point where the Earth's center was located exactly 365 days ago. On this view, there is a fact regarding the distance between two events that are not simultaneous: each is located at a certain point of space, and there is a fact regarding the distance between these points.

On Newton's view, although there is a fact regarding a body's velocity, this fact is undiscoverable: there is no way to tell whether the origin of a given inertial frame is moving uniformly at (say) 5 cm/s, or instead moving uniformly at 10 cm/s, or instead at rest. That is because (to repeat) a body's absolute velocity makes no difference to anything else; the laws of Newtonian mechanics reveal no observable phenomenon that would be different if all bodies at every moment were moving with (say) an additional 5 cm/s in a common direction. The bodies' relative positions, for example, would be unchanged. But if absolute velocities play no part in the Newtonian mechanical explanations of our observations, how could our observations justify our believing that bodies possess absolute velocities?

This line of thought, though failing to rise to the level of a conclusive argument, motivates the rival view to Newton's, according to which a body has no absolute velocity. On this view (sometimes called "Galilean" or "neo-Newtonian" space-time), there *is* a fact about whether or not the body is accelerating, but its acceleration cannot be defined as its real velocity's rate of change. Rather, its acceleration must be defined as the rate of change of its velocity relative to any inertial frame. Although the body's velocity is frame-dependent, the velocity changes at the *same* rate in every inertial frame, since those frames are not accelerating relative to one another. The objective facts, on which all inertial frames agree, include one body's velocity relative to another and the time interval between (or simultaneity of) two events. But there is no objective fact regarding the distance between two non-simultaneous events: if every point at rest in system B is moving at a steady 5 cm/s to the right relative to system A, then if two events occur 1 second and 5 centimeters apart according to A, they may both occur at the same point according to B. So there is no fact about whether or not two non-simultaneous events occurred at the same location.

In chapter 5, we explored the possibility of interpreting energy as real stuff stored by, or a real property possessed by, a body or a field, thereby securing the reality of fields and thus making electromagnetic influences obey spatiotemporal locality. With the aid of the Poynting vector, we can depict energy as flowing through the field, and although we have seen no conclusive argument for this interpretation, it seems attractive. In the words of Poynting and J. J. Thomson:

> There is something which we term "energy," and which may be recognised in various forms. When it disappears in one form it appears in one or more other forms. . . . The belief in the identity of energy is no doubt metaphysical, as metaphysical as is our belief in the continued existence of any portion of matter, and its identity under various modifications. As, however, the metaphysical addition somewhat simplifies the form of the statement, and is never likely to lead us wrong in our experimental interpretation, we see no reason to exclude it. Were we to do so we should have to speak of the correlation of the energies, not of the constancy of energy. Instead of describing the conversion of, say, kinetic energy into heat, we should have to say that kinetic energy disappeared, and that at the same time heat energy appeared.   (1928: 110)

However, according to Galilean space-time, energy cannot be real stuff or even one of a system's objective properties. For if there is no fact regarding a body's velocity (v), then there is no fact regarding its kinetic energy ($\frac{1}{2}\,mv^2$); energy is not an invariant. This seems to have been one of Heaviside's principal reasons for disagreeing with Lodge's conception of energy as real, its parcels possessing their own identities like bits of matter. In Galilean space-time, some bit could exist only in a frame-dependent way, not objectively. Heaviside wrote:

> [W]e need not go so far as to assume the objectivity of energy. This is an exceedingly difficult notion, and seems to be rendered inadmissible by the mere fact of the relativity of motion, on which kinetic energy depends. We cannot, therefore, definitely individualize energy in the same way as is done with matter.   (1925: II, 525–6, also 521; cf. 1922: I, 75)

Strangely, Heaviside did not seem to recognize that the same argument impugns not only the "thingness" of energy but also energy's status as a real property of bodies and fields.[2]

That a body has a velocity only relative to some inertial frame (as Galilean space-time says) is what Heaviside calls "the relativity of motion." Let us see how this doctrine undermines the reality of energy as stuff that flows or a property that can be treated as "flowing." Imagine two icy comets, each traveling inertially, which collide, stick together, and then melt. In the reference frame from which one comet is at rest (and so has zero kinetic energy), the moving comet's kinetic energy is converted by the collision into heat energy, which then flows into the other comet, melting it. The direction of energy flow is thus opposite from each comet's viewpoint; each sees the flow as into itself from the other comet. So the path of energy flow is not invariant, and hence not real, if all inertial frames are accurate in describing the universe.[3]

Of course, Galilean space-time denies momentum's reality just as it denies energy's reality: both energy and momentum depend on velocity, and according to Galilean space-time, a body's velocity has no objective reality. But if neither energy nor momentum is real, then we cannot argue for the electromagnetic field's reality from the fact that it possesses energy and momentum (as I explained in chapter 5). And without the field, we cannot portray electromagnetic phenomena as having local causes.

2   Relativistic Invariants and the Unification that Relativity Achieves: Space and Time

Unlike Newton's physics, the laws of classical electromagnetic theory refer to velocities. For example, $\boldsymbol{F} = (q/c)\ \boldsymbol{v} \times \boldsymbol{B}$ (equation 2.7) refers to the velocity $\boldsymbol{v}$ of a body with charge q that feels magnetic force $\boldsymbol{F}$

exerted by magnetic field **B**. The speed of light c appears not just in this law, but also in connection with Maxwell's equations as the speed at which electromagnetic influences travel across empty space (like ripples in a pond). A velocity appearing in these laws could be interpreted as the body's *real* velocity (that is, its velocity relative to the points of absolute space), in which case Galilean space-time must be false. Alternatively, a nineteenth-century physicist might have interpreted a velocity figuring in these laws as the body's velocity relative to the aether (recall box 2.2). This interpretation came in two versions. On the one hand, the aether might itself have a certain absolute velocity – that is, a certain velocity relative to absolute space. On the other hand, perhaps neither the aether nor any body has an objective velocity, in accordance with Galilean space-time.

On *any* of these views, we should on different occasions measure light as traveling at different speeds relative to ourselves. Here's why. Suppose the light is traveling to the right at c: 300,000,000 m/s. (The velocities in this example could all be objective or merely relative to the aether.) Suppose we are traveling to the *left* at 50 percent of c. Then we should measure the light as traveling to the right at 150 percent of c. On the other hand, if we are traveling to the *right* at 50 percent of c, then relative to us, the light is moving to the right at 50 percent of c. In the 1860s, Maxwell conducted experiments intended to measure differences in the speed of light relative to different reference frames; he failed to detect any differences but was uncertain of what conclusion to draw from this result (Maxwell 1890: II, 769–70). Various later, more careful experiments (the most famous carried out in Cleveland, Ohio, in the 1880s by Albert Michelson and Edward Morley) likewise revealed no variation in light's measured speed in any inertial reference frame. Furthermore, only in a frame at rest relative to the aether should light move in all directions with the same speed, yet in all of the reference frames in which Newton's laws hold classically, light is measured to have the same speed in any direction. Here, then, is an invariant that cannot be accounted for by any of the interpretations I have mentioned. A moving car appears to be going slower if you are running in the direction it is moving and faster if you are running in the opposite direction. Strangely, the same doesn't apply to a light beam: no matter how fast you run after it, it always escapes from you at the same speed.

Remarkably, Einstein at age 16 seems to have taken this invariant for granted (Einstein 1949: 53). In his 1905 paper, he combined this invariant with the "principle of relativity": that any genuine law of physics must hold in every inertial reference frame – that is, in every frame in which classical mechanics says that Newton's "laws of motion" hold. (Newton's "laws" *quote-unquote* because according to relativity theory, they are not genuine laws, since they are merely approximately true under certain circumstances.) The principle of relativity is one expression of the more general principle that reality can be described accurately from any inertial frame: the natural laws are genuine features of reality, so they must hold in every inertial frame.

Suppose, in particular, that Maxwell's equations are natural laws. The principle of relativity then entails that Maxwell's equations are obeyed by the phenomena as measured from any inertial frame. But what does this mean? Maxwell's equations refer to the speed of light (presuming light to consist of waves of disturbance to the electric and magnetic fields), but as we have seen, it is not obvious how this speed should be interpreted. As light's absolute speed? As its speed relative to the aether? Notice that the principle of relativity is satisfied by the natural laws' dictating in every inertial frame the speed of sound relative to the air through which the sound travels. In different inertial frames, the same sound wave travels at different speeds, but so does the air through which it is traveling, so the sound's speed relative to that air is the same. Likewise, if light had the same speed relative to the aether in every inertial frame, then the principle of relativity would have been satisfied. However, as I just mentioned, experiments reveal that light (unlike sound) travels ·at the same speed in every inertial frame. So in order for Maxwell's equations to hold in every inertial frame, the speed they assign to light cannot be interpreted as its speed relative to the aether.

From the principle of relativity (or, perhaps, the more general principle that the universe can be described accurately from every inertial frame) and the law that the speed of light is the same in every inertial frame (along with other laws of nature, such as $F = (q/c)\ v \times B$), there follow all of the famous consequences of relativity theory. Since light has the same speed c (through a vacuum) in any inertial frame, we do not need to distinguish a special reference frame relative to which c is the speed of light. Hence, we do not need to hypothesize the aether as the thing to which this special frame is attached. Since two inertial frames moving uniformly relative to each other disagree about which of them is moving and which is at rest, but the universe can be described with equal accuracy from either frame, it follows that there is no fact about which frame is *really* moving. Galilean space-time was correct: there are no absolute velocities, and so no real quantities of energy and momentum. Thus, we cannot use the ontological status of energy and momentum to underwrite the reality of fields. We have lost our best argument so far for spatiotemporal locality!

But don't worry. Though energy and momentum are not invariant quantities, we will shortly encounter another invariant: mass. By the end of this chapter, we will have used mass's ontological status to underwrite the fields' reality, and thus spatiotemporal locality.

But we must get there one step at a time. Let's begin by examining one of relativity's most notorious consequences. Suppose we have an apparatus involving two mirrors facing each other, with light bouncing back and forth in between. Consider an inertial frame in which the apparatus is at rest. If d is the distance between the mirrors, then a round trip is $2d$, and so light covers this distance in $t = 2d/c$ seconds. Now consider an inertial frame relative to which the apparatus is moving at speed v in a direction perpendicular to the line between the mirrors (figure 8.2). In this frame, light's path is like the path that would be taken, relative to you, by a ball you were watching being tossed back and forth between the occupants in the rear seat of a car, as the car was traveling at speed v in a direction perpendicular to the rear seat (recall figure 8.1). The ball goes on a diagonal path, and

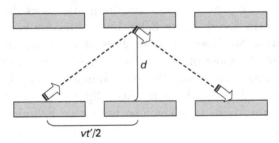

Figure 8.2   The apparatus (consisting of two mirrors facing each other) at three moments, as seen from a reference frame relative to which the apparatus travels to the right. The light (white arrows) appears to travel in a path shaped like an upside-down letter V, each side of which is length x. Let v be the speed at which the apparatus is moving, relative to this reference frame. Let t' be the time it takes for the light to complete a round trip between the mirrors. Then during one leg of the trip, the apparatus moves a distance $vt'/2$. If d is the distance between the two mirrors, then a right triangle is formed with hypotenuse x and legs d and $vt'/2$. By the Pythagorean theorem, $d^2 + (vt'/2)^2 = x^2$.

similarly for the light in our example. So the light's round-trip path is longer than $2d$; by the Pythagorean theorem, it is $2\sqrt{(d^2 + v^2 t'^2/4)}$, where t' is the time it takes in this frame for light to complete a round trip.[4] Since the speed of light is again c in this frame, $t' = (2/c)\sqrt{(d^2 + v^2 t'^2/4)}$, which is plainly greater than $t = 2d/c$. Specifically:

$$t'^2 = 4/c^2[d^2 + v^2 t'^2/4] = t^2 + v^2 t'^2/c^2.$$

So

$$t' = t/\sqrt{[1 - (v^2/c^2)]} = \gamma t,$$

where γ (gamma) is an abbreviation for $1/\sqrt{[1 - (v^2/c^2)]}$. That $t' > t$ means that the two frames disagree on the interval of time between the same two events: in the frame attached to the apparatus, t seconds have passed, while in the frame relative to which the apparatus is moving, t' seconds have passed. According to relativity theory, the universe can be accurately described from each of these frames; only what is the same in every inertial frame can be a genuine feature of reality. So there is no objective fact regarding the interval of time between two events; time intervals (and spatial separations as well) are frame-dependent.

Textbook discussions of relativity typically include thought experiments like the following: Imagine that someone is standing in the middle of a train, and someone on the ground sees the train pass by at a uniform speed (figure 8.3a); both observers are attached to inertial frames. Suppose that both ends of the train are struck by lightning bolts. Suppose that according to the observer on the ground, the two lighting bolts struck the train simultaneously: light reaches her simultaneously from the two strikes (figures 8.3b and 8.3c). The observer on the train, however, concludes that lightning struck the front of the train before the rear. She draws this conclusion from the fact that light from the front strike reaches her before light from the rear strike.

This kind of thought experiment does *not* show that according to common sense, the simultaneity of two events is not an objective fact. According to common sense, at least one of the observers in this example is mistaken in taking the order of arrival of light from the lightning strikes as directly revealing the order in which those strikes occurred. Presumably, the observer on the train is moving toward the

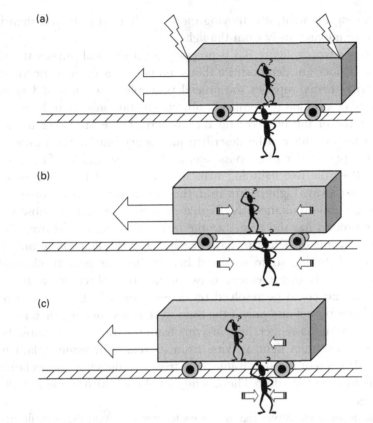

Figure 8.3   In (a), lightning bolts simultaneously strike both ends of a train moving uniformly to the left. In (b), beams of light from the two lightning strikes travel toward observers on the train and on the ground. We are observing the scene from a reference frame attached to the ground. In (c), the light beams from the two strikes simultaneously reach the observer on the ground. However, because the train is moving to the left, the beam from the lightning strike on the right has not yet caught up to the observer on the train.

light coming from the train's front and away from the light coming from the train's rear. (It *is* a train, after all.) Therefore, the light coming from the front is approaching the train passenger more quickly than the light coming from the rear, throwing off her determination of the order in which the strikes occurred.

This commonsense response directly conflicts with the law that light travels at the same speed in any inertial frame. Of course, this principle is utterly contrary to common sense, which would say that if you are chasing the light, it is moving more slowly relative to you than if you were moving away from the light.

In other words, figure 8.3 is precisely what classical physics would tell us. It does not demonstrate that simultaneity is frame-dependent. Relativity theory supplies the crucial principle that figure 8.3 is an accurate description of events not objectively, but only from the reference frame of the observer on the ground. There are other frames from which events can be described just as accurately. For example, from the perspective of the passenger at the precise middle of the train (figure 8.4), the two lightning strikes (at the ends of the train) occur equally far away. Light beams from the two strikes reach the observer on the ground simultaneously. By that moment, however, the observer on the ground has slid nearer to the train's rear than to its front. So the two beams of light can have reached her simultaneously only if the front lightning strike occurred before the rear one. In classical mechanics, this disagreement between the two observers is to be expected and could be resolved by determining which observer was really in motion: if and only if the train were really moving, then relative to the train passenger, light arriving from the train's front would be moving faster than light arriving from its rear. But again, relativity theory denies that in either frame, light from one direction moves faster than light from the other. There is no fact about which frame is really moving.

The point of examples like this is *not* to persuade you that simultaneity is frame-dependent *simply by appealing to your common sense*. As an argument for relativity theory, it is utterly circular. The example *presupposes* relativity when it appeals to the principle that there is no fact about whether a given inertial frame is really moving. *Since* there is no fact regarding whether the train is moving uniformly or at rest, the train passenger cannot be advised to take her motion into account in determining which lightning strike occurred first. *Since* the universe can be accurately described from any inertial frame and the example supplies inertial frames that disagree on the simultaneity of the lightning strikes, relativity theory entails that there is no fact about whether two events are simultaneous. That entailment is what the example demonstrates.

As I pointed out in chapter 1, Newtonian gravity seems to involve instantaneous action at a distance, in violation of spatial locality. The cause (in the Sun: a body's possessing a certain mass) is simultaneous with its distant effect (on Earth: a body's feeling a certain gravitational

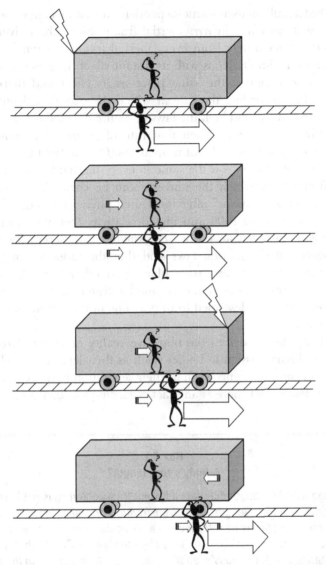

Figure 8.4   The events as described from a reference frame moving with the train. From this vantage point, the train is stationary but the observer on the ground is moving to the right. Notice that the reference frame on the ground (figure 8.3) and the reference frame moving with the train (this figure) agree that the train passenger sees the light beam coming from the front of the train before she sees the light beam coming from the train's rear, whereas the ground-based observer sees the two light beams simultaneously. But the two reference frames disagree on the relation between the times of the lightning strikes.

force). But simultaneity is frame-dependent; the cause cannot be simultaneous with its effect in *every* inertial frame. Since the universe can be described accurately from every inertial frame, it cannot *really* be that a cause is simultaneous with its distant effect. In general, suppose that a cause is not at the same place as its effect and there is an objective fact regarding the rate at which the cause's influence propagates, as there would be if the laws of nature deemed this velocity to be a universal constant. Then that rate of propagation must be *c*, since this is the only speed that is invariant. (A quantity that – according to relativity theory – is the same in every inertial frame, and so in every frame from which the universe can be correctly described, is called "Lorentz-invariant" after the Dutch physicist Hendrik Lorentz (1853–1928), who in 1904 anticipated relativity theory by identifying the relations between an event's coordinates in different frames.) In other words, if the natural laws said that the influence propagates instantaneously (or at any particular speed other than *c*), then the natural laws could not hold in every inertial frame, and so the universe could be accurately described from only one of these frames, contrary to relativity.

Relativity theory denies the objective reality of various properties that we ordinarily assign to bodies (such as their length and velocity – see box 8.1) and to events (such as their separation in space and time), just as it denies that there is any fact regarding whether a given force

---

Box 8.1
A body's real length?

Though a body's length differs in different frames, and so is not Lorentz-invariant, a body's length *in a given frame* is the same in all frames. (This quantity carries its reference to a particular frame along with it, so to speak.) Therefore, this quantity is objectively real. One might then be tempted to say that a body's length *relative to a frame in which the body is at rest* (a Lorentz-invariant quantity) is its *real* length. But why should we take this frame, rather than any other, as giving the body's *real* length? Granted, it is a frame picked out by the body, but why is that a good reason? (Admittedly, the body's length relative to this frame is important for practical purposes, since it is approximately equal to all of our measurements of the body's length when the body's speed relative to us remains small compared to the speed of light.)

is really electric or magnetic (as we saw in chapter 7). Recall from the start of this section that the laws of classical electromagnetic theory refer to velocities, as in $F = (q/c) \, v \times B$. According to relativity theory, v is not the *real* speed of the body – since there are no absolute speeds. Rather, v is its speed in the inertial reference frame in which we are working – the frame in which the magnetic field takes on the value B that we are using. In contrast, c is the same in every inertial frame, as is q. The equations of classical electromagnetic theory hold in *any* inertial frame, not just in one that is really at rest (or at rest relative to the aether).

One quantity that *does* turn out to be Lorentz-invariant is the "spatiotemporal interval" I between two events. If Δs is the distance between two events relative to a certain inertial frame and Δt is the span of time separating them in that frame, then $I = (c\Delta t)^2 - (\Delta s)^2$. The interval I between two events is a real feature of the universe, though in different inertial frames, the same I is broken down into different spatial and temporal components. No decomposition is more accurate than any other, just as in different inertial frames, the electromagnetic field appears as different combinations of E and B. The interval $I = (c\Delta t)^2 - (\Delta s)^2$, with its temporal and spatial components, looks very much like the electromagnetic invariant $E^2 - B^2$, with its electric and magnetic components. Every decomposition of the spatiotemporal interval between two events into Δt and Δs, or of the electromagnetic field into E and B, reflects not just reality, but also a particular reference frame's own perspective. The interval I depends only on how the universe really is, uncontaminated by any contribution from us in describing the universe.

That the span of time between two events is frame-dependent might lead you to worry that for a cause C and its effect E, there is an inertial frame in which C comes *after* E. This might reasonably lead you to doubt whether C's being a cause of E is an objective fact. But there is no cause for worry so long as in some inertial frame, light beginning at C would have enough time to reach E's location before E occurs there – that is, so long as $I > 0$. (I is then termed "timelike.") In that case, since I is invariant, every inertial frame agrees that E occurs after C; the order of C and E in time is invariant. (Their arrangement in space is *not* invariant; there is even an inertial frame in which C and E occur at the same location, though of course at different times.) Likewise, suppose that in some inertial frame, light leaving C would reach E's location just as E occurs there. That is, suppose $I = 0$. (I is then termed "lightlike.") In that case, once again, no inertial frame says that E occurs before C. Therefore, as long as causal influences in a given inertial frame propagate no faster than c, the time order of C and E in one inertial frame is not reversed in any other inertial frame. Therefore, we can still regard C's causing E as an objective fact, and so we can continue to ask whether causal relations obey spatiotemporal locality. The book can continue!

For example, suppose that in a given inertial frame, a body's having a certain charge and moving with velocity v at space-time point C is a cause of the fields at distant space-time point E. Now take a different inertial frame, in which spatiotemporal locations C′ and E′ correspond to locations C and E, respectively, in the original frame. In the new frame, the values of the fields at E′ may differ from their values at E in the original frame, and the charged body's velocity at C′ may differ from v. However, the speed c at which causal "news" travels is the same in both frames. Moreover, since the spatiotemporal interval is Lorentz-invariant, events separated by $I = 0$ in one inertial frame (so that light leaving one arrives at the other) are separated by $I = 0$ in every inertial frame. Hence, in the new frame, causal news leaving C′ arrives at E′; the frames agree on which events help to cause which other events (Rosser 1997: 373–5).

To sum up this section: Relativity theory *unifies* space and time in the same sense of "unity" that I explained (in chapter 7) in connection with E and B: space and time are just different ways in which the same real thing ("space-time") appears from a given perspective. As Hermann Minkowski (Russian-German mathematician, 1864–1909) famously put it:

> Henceforth space by itself, and time by itself, are doomed to fade away into mere shadows, and only a kind of union of the two will preserve an independent reality.   (1952: 75)

In chapter 1, I defined "spatiotemporal locality" in terms of the distance in space and separation in time between an event and its causes. We now see that these distances and separations are not objective matters of fact. However, whether or not spatiotemporal locality is satisfied is supposed to be an objective matter of fact. The simplest way to ensure this is to specify that for spatiotemporal locality really to apply, my chapter 1 definition of "spatiotemporal locality" must be satisfied *in every inertial frame*.

3 Relativistic Invariants and the Unification that Relativity Achieves: Energy and Momentum

In chapter 5, I explained that in classical physics, there are two ways to think about the accumulation of force. We can consider the accumulated force felt by a body over a given span of time ($\int \boldsymbol{F} dt$), which equals the net change in the body's momentum ($\Delta \boldsymbol{p}$). Alternatively, we can consider the accumulated force felt by a body as it moves across a certain distance ($\int \boldsymbol{F} \cdot \boldsymbol{ds}$), which equals the net change in the body's energy. Momentum and energy are, of course, the two classically conserved quantities that were crucial to the argument (given in chapter 5) for the reality of fields, supporting spatiotemporal locality. Since neither a span of time nor a distance is Lorentz-invariant, neither is real according to relativity theory, and so we should expect neither momentum (force accumulated over time) nor energy (force accumulated over space) to be real. In section 1, we saw that this was the case for "energy" and "momentum" as defined in classical physics; the same will be true in relativity, as we shall see in a moment. This conclusion appears to undermine our argument for the reality of fields from the need for something to hold certain quantities of momentum and energy.

On the other hand, as we have seen, there *is* something real (Lorentz-invariant) that appears in a given inertial frame as a certain span of time coupled with a certain distance separating two events, and that appears in a different inertial frame as different separations in distance and time between those events. This objective fact is the spatiotemporal interval I between the events. Analogously, we might expect there to be *some* objective fact that appears as different combinations of momentum and energy in different inertial frames. If so, relativity would thereby unify the two senses of accumulated force in classical physics, revealing them to be different components of the same real quantity. (This invariant quantity might even turn out to be useful, in place of energy and momentum, for underwriting the reality of fields, and hence spatiotemporal locality!)

These expectations turn out to be met. A body's momentum $\boldsymbol{p}$ in classical physics is $m\boldsymbol{v}$, where m is the body's mass and $\boldsymbol{v}$ is its velocity. In a given coordinate system, there are three axes ($x, y,$ and z – one for each dimension of space), and so $\boldsymbol{p}$ has three components ($p_x, p_y,$ and p_z) corresponding to the three components of a body's velocity $\boldsymbol{v}$: the rates of change of its $x, y,$ and z coordinates ($v_x, v_y,$ and v_z). The body's speed v equals $\sqrt{[v_x^2 + v_y^2 + v_z^2]}$. But in relativity theory, as we have seen, a body's spatial coordinates in a given inertial frame present an incomplete picture of something real. To capture an objective fact, a quantity must include not just spatial components, but the relevant time component as well. Accordingly, a body's "relativistic velocity" V contains *four* components – which, like I's components, are frame-dependent. The first three components of V (the body's velocity "4-vector") are the rates at which each of its three spatial components in the given frame are changing relative to time in that frame (v_x, v_y, and v_z) multiplied by the rate at which time in that frame is passing relative to the rate at which time is passing in a frame in which the body is at rest (which, as we saw earlier, equals γ). So these three components equal γv_x, γv_y, and γv_z. The fourth component is simply γc. Just as the frame-dependent quantities Δs and Δt combine to form the invariant I, so the frame-dependent quantities γv and γc form an invariant, c^2:

$$(\gamma c)^2 - (\gamma v)^2 = \gamma^2[c^2 - v^2] =$$
$$(1/[1 - (v^2/c^2)])c^2[1 - (v^2/c^2)] = c^2. \qquad (8.1)$$

Just as classically, a body's momentum is $m\boldsymbol{v}$, so the corresponding relativistic 4-vector is mV. (Here m is what is sometimes called the body's "rest mass" – though, as I shall explain shortly, I prefer simply to call it the body's "mass.") The spatial components of this 4-vector are $m\gamma v_x$, $m\gamma v_y$, and $m\gamma v_z$, which in relativity are defined as the components of the body's "momentum" $\boldsymbol{p}$, amending the classical notion of momentum $m\boldsymbol{v}$.

To further extend the analogy with classical physics, $m\gamma$ is sometimes called the body's "relativistic mass," so that momentum remains mass times velocity – but now "relativistic mass." Whereas m is invariant, $m\gamma$ is not, since γ is obviously frame-dependent (as it is a function of v). I shall reserve the term "mass" for m and say more in the next section about why I don't want to think of $m\gamma$ as "relativistic mass."

The time component of mV is $m\gamma c$, which in relativity is defined as the body's "energy" $\mathcal{E}$ divided by c. In other words, $\mathcal{E} = m\gamma c^2$. (So $\mathcal{E}$ equals "relativistic mass" times c^2. This is one version of Einstein's formula "$\mathcal{E} = $ (mass) c^2." In the next section, I shall say a good deal more about this famous equation.) As in classical physics, the force accumulated over the distance covered in moving the body from one location to another (that is to say, the "work" done) is the difference between the body's final energy and its initial energy.[5] The 4-vector $\mathbf{P}$, then, has components p_x, p_y, p_z, and $\mathcal{E}/c$.

The payoff from all this effort is that the "energy-momentum 4-vector" $\mathbf{P}$ *unifies* energy and momentum in the following sense. In any inertial frame, a closed system's total energy and total momentum (as defined relativistically) are each "conserved." That is, in any inertial frame, the system's total energy does not change as time passes, and likewise for its total momentum. But these totals differ in different inertial frames. In other words, total energy and total momentum are frame-dependent and therefore not real. However, energy and momentum *when taken together* – that is, the components of the energy-momentum 4-vector – form a Lorentz-invariant quantity, m^2c^2:

$$(\mathcal{E}/c)^2 - p^2 = (m\gamma c)^2 - (m\gamma v)^2 = m^2[(\gamma c)^2 - (\gamma v)^2] = m^2 c^2, \qquad (8.2)$$

where the final step uses equation (8.1). The equation $m^2c^2 = (\mathcal{E}/c)^2 - p^2$, decomposing the invariant m^2c^2 into energy and momentum components, looks a great deal like the spatiotemporal interval I expressed in terms of temporal and spatial components ($I = (c\Delta t)^2 - (\Delta s)^2$) and also looks like the electromagnetic invariant $E^2 - B^2$, with its electric and magnetic components.

The quantity m^2c^2 is Lorentz-invariant because m and c are Lorentz-invariant. But the decomposition of m^2c^2 into energy and momentum is frame-dependent, just as in different frames, the electromagnetic field appears as different combinations of electric and magnetic fields and the spatiotemporal interval I between two events appears as different combinations of temporal and spatial separations. A closed system's total mass m is not only invariant (and so a real characteristic of the system), but also conserved in any inertial frame: both $\mathcal{E}$ and p are conserved, and by equation (8.2), m is determined by $\mathcal{E}$ and p. All of this applies only to *bodies*, that is, bits of ordinary matter. The relativistic expressions for the *field*'s energy and momentum are identical to those in classical electromagnetism since, as I mentioned in chapter 2, classical electromagnetic theory was already relativistic; relativity theory is, in effect, trying to bring the rest of physics into line with classical electromagnetism.[6]

Let me elaborate the sense in which relativity theory unifies energy and momentum. In classical physics, both energy and momentum are conserved, as is mass. In fact, these conservation laws are not independent of one another in classical physics: momentum conservation follows from energy conservation and mass conservation (von Laue 1949: 516). But, as we know from chapter 7, even such a tight link between momentum, energy, and mass does not mean that the three concepts have been *unified* in the deepest sense. That unity is what relativity theory supplies: that momentum and energy are nothing but different aspects of the same real thing (mass), just as $\boldsymbol{E}$ and $\boldsymbol{B}$ are frame-dependent aspects of another real thing (the electromagnetic field), and Δt and Δs are frame-dependent aspects of still another real quantity (the spatiotemporal interval). Of course, relativity theory not only upholds energy, momentum, and mass conservation, but also allows any two of these conservation laws to entail the third, since a system's total energy, momentum, and mass are related by equation (8.2).[7] But again, this entailment is not how relativity *unifies* energy, momentum, and mass. Rather, the key point is that neither energy nor momentum is Lorentz-invariant, but together they form a Lorentz-invariant quantity: mass. We must now try to understand mass better.

4 Mass and the Meaning of "$\mathcal{E} = mc^2$"

It is sometimes said that according to relativity theory, energy is "equivalent" to mass in virtue of the famous equation "$\mathcal{E} = mc^2$." This equation holds if "m" stands for a body's "relativistic mass," as I mentioned earlier. A body's energy ($m\gamma c^2$) and "relativistic mass" ($m\gamma$) are plainly just two ways of expressing the same property, differing merely in units. But this property is *not* objectively real, since it is not Lorentz-invariant. When we change reference frames, a body's "relativistic mass" changes, but of course, this "change" does not reflect anything happening to the body. A body's γ increases with its v, so a body's "relativistic mass" increases with its speed, approaching infinity as the body's speed approaches c.

In contrast, a body's m is independent of its speed. Unless I say otherwise, I will take "m" in "$\mathcal{E} = mc^2$" to be the body's mass – the real, Lorentz-invariant quantity. Then "$\mathcal{E} = mc^2$" holds only when the body is at rest (that is, when $\boldsymbol{p} = 0$); $\mathcal{E} = mc^2$ then follows immediately from $(\mathcal{E}/c)^2 - p^2 = m^2 c^2$, which was equation (8.2).

When the body is at rest, its $\gamma = 1$, and so its "relativistic mass" $m\gamma$ equals its mass m. That is why "m" is sometimes called the body's "rest mass." Because "relativistic mass" is not an invariant quantity, the best thing to do in order to avoid confusing frame-dependent quantities with invariant ones is just to avoid using the term "relativistic mass." We must then also abandon the term "rest mass," since this term makes sense only in view of the term "relativistic mass": the body's "rest mass" is its "relativistic mass" when at rest. We should instead use just the term "mass," applying it to the invariant quantity symbolized m.

Remember that the only reason for referring to $m\gamma$ as a kind of mass is to allow $\boldsymbol{p} = m\gamma\boldsymbol{v}$ to take the form "momentum = mass times velocity," as is the case in classical physics (where $\boldsymbol{p} = m\boldsymbol{v}$). But this mnemonic device should not lead you to think of a body's "relativistic mass" as anything like the amount of matter of which the body is made. Bear in mind that there are many similar mnemonic devices. For example, relativity says that when a force $\boldsymbol{F}$ acts parallel to a body's velocity, then the body's acceleration $\boldsymbol{a}$ is given by $\boldsymbol{F} = m\gamma^3\boldsymbol{a}$. Since classically, $\boldsymbol{F} = m\boldsymbol{a}$, $m\gamma^3$ is sometimes called a body's "longitudinal mass." But "longitudinal mass" no more measures a body's quantity of matter than "relativistic mass" does. Furthermore, neither relativistic mass nor longitudinal mass is Lorentz-invariant, so neither stands on an ontological par with m. That Einstein did not regard "rest mass" and "relativistic mass" as standing on the same ontological footing is evident from his many remarks urging others to introduce as "mass" only a quantity independent of motion (Fine 1977: 538; Okun 1989).

Let us look more closely at $\mathcal{E} = mc^2$. You might have wondered how this equation can hold, considering that m and c are the same in every inertial frame but $\mathcal{E}$ is frame-dependent. We have just seen the answer: $\mathcal{E} = mc^2$ holds only for $\mathcal{E}$ in the frame where $\boldsymbol{p} = 0$.

Sometimes, physics textbooks present examples in which mass is "converted" into energy, and the equation $\mathcal{E} = mc^2$ is then used to determine how much energy is "equivalent" to (at least in the sense that it replaces) the mass that disappears. For example, after a radioactive nucleus decays, there is a "mass defect": the sum of the masses of the daughter bodies is less (by Δm) than the mass of the original nucleus. Some of the original mass (Δm) is said to have been "converted" into the kinetic energy $\mathcal{E}$ of the daughter bodies, where $\mathcal{E} = (\Delta m)c^2$. (Since c is so large, a very small mass can be "turned into" a great deal of energy.) For instance, when a tritium nucleus (one proton, two neutrons) decays into a helium-3 nucleus (two protons, one neutron) along with an electron and an anti-neutrino, the tritium's mass exceeds the sum of the products' masses by a small quantity that is "equivalent" to about 0.000 000 03 ergs of energy.

But if the interconvertibility of mass and energy is all that is meant by mass–energy "equivalence," then this "equivalence" does not mean that energy really *is* mass. It means only that when a certain amount of mass Δm disappears, the *equivalent* amount of energy (the amount for which it is exchanged, the amount that takes its place) is $(\Delta m)c^2$. This does not show that mass is the same thing as energy or even that the mass gets *turned into* energy – only that when one disappears, the other replaces it. Yet textbooks commonly include passages like this:

The equation $[\mathcal{E}] = mc^2$ is perhaps the most famous equation of twentieth-century physics. It is a statement that mass and energy are two forms of the same thing, and that one can be converted into the other.   (Kane and Sternheim 1978: 493)

At least this passage distinguishes the unity of mass and energy from their interconvertibility. Sometimes, authors try to slide from one to the other:

[S]ince mass and energy are totally convertible into each other, they are really just different names for the same concept.   (Thorne 1994: 441)

As we have seen, the conversion of an electron-positron pair, for example, into gamma radiation or its mirror phenomenon is an incontestable experimental confirmation of the assertion of the theory of relativity that mass and energy are mutually and completely interconvertible. This state of affairs raises the following questions: Are not the two entities which are interchangeable essentially the same? Is not what is generally spoken of as an equivalence in reality an identity? Are therefore not "mass" and "energy" merely synonyms for the same physical reality, which . . . may perhaps be termed "massergy"? (Jammer 1961: 184)

The author clearly intends the answers all to be "yes."

But from the fact that one thing can be converted into the other, it simply does not follow that the two are actually different forms of the same thing.[8] James Prescott Joule (English physicist, 1818–89), for instance, famously suggested that all forms of energy (heat, kinetic energy, gravitational potential energy, chemical bond energy, electrical energy, and so on) are interconvertible, but he did not go so far as to hold that they are all different forms of the same thing. In 1843, he showed experimentally that (what he called) the "mechanical equivalent of heat" is (in modern units) 4.18 joules of energy per calorie of heat. In other words, Joule demonstrated that a 1 kg mass falling through a distance of 42.67 cm contributes exactly enough mechanical energy (in doing work by, say, turning a stirring paddle immersed in water) to add 1 calorie of heat to the water. But he saw this "equivalence" as involving how much of one thing replaces a given quantity of the other – their fixed "rate of exchange." He did not understand the mechanical equivalent of heat as we do today: as a conversion factor, that is to say, a way to change our units for measuring the same thing, because heat is just a particular form of energy (Elkana 1974; compare Einstein 1989: 386).

Yet this is how "mass–energy equivalence" is often understood:

> [J]oules and kilograms are two units – different only because of historical accident – for one and the same kind of quantity, mass-energy. . . . The conversion factor c^2, like the factor of conversion from . . . miles to feet, can today be counted, if one wishes, as a detail of convention, rather than as a deep new principle.   (Taylor and Wheeler 1966: 137)

As Einstein himself writes:

> Mass and energy are therefore essentially alike; they are only different expressions for the same thing.   (1953: 45)

But how is mass–energy equivalence in this sense to be understood?

For instance, if mass is a real property (since it is Lorentz-invariant) whereas energy is not, then how can mass and energy measure the same thing ("mass–energy") *in the same sense* that distance in miles measures the same thing as distance in feet? The distance in miles between New York and Boston is neither more nor less real than the distance in feet! Moreover, in what sense can mass be *converted* into energy when mass and energy are not on a par in terms of their reality? When mass turns into energy, is it thereby ceasing to be real, since energy is frame-dependent?

Here is another puzzle. Take a case where mass allegedly turns into energy, such as the decay of a radioactive nucleus. There is a "mass defect" and so, apparently, mass fails to be conserved. Yet mass *conservation* must hold in relativity theory, since $\mathcal{E}$ and p are conserved and together determine m by equation (8.2). How can "missing mass" be reconciled with mass conservation? For that matter, energy is also conserved in relativity theory. If some mass gets turned into energy, isn't there more energy than before?

Furthermore, if mass occasionally turns into energy (as when a tritium nucleus turns into a helium-3 nucleus and other things), then how can joules and kilograms be like miles and feet? Miles never *turn into* feet (except in the sense that when we do a calculation, we sometimes change the units we are using). A "conversion" of miles into feet is not a *physical* transformation, like the metamorphosis of a caterpillar into a butterfly. Is the conversion of mass into energy (and vice versa) a physical transformation? It appears to be, if we think of it as part of the decay of a tritium nucleus. That decay is a physical transformation, and mass apparently gets turned into energy when some of the *stuff* of which the tritium nucleus is made gets transformed into the *motion* of the daughter bodies.

On the basis of $\mathcal{E} = mc^2$ (when $\boldsymbol{p} = 0$), Einstein (1989: 174; also 1931: 55–6) says that "the mass of a body is a measure of its energy content [in the $\boldsymbol{p} = 0$ frame]." Yet this remark may mislead us by suggesting that energy content is an objective fact whereas mass is merely some sort of indicator, symptom, or manifestation of energy. In fact, however, it is mass rather than energy that is Lorentz-invariant.[9]

The reality of fields, and hence spatiotemporal locality, will turn on properly interpreting mass in relativity theory. We must do so without taking refuge in remarks like these:

> [W]e must abandon attempts to say what energy is. We must say simply: there is something quantitative, to which we give the name "energy"; this something is very unevenly distributed in space; there are small regions in which there is a great deal of it, which we call "atoms", and are those in which, according to older conceptions, there was matter. . . . Mass is only a form of energy. . . . It is energy, not matter, that is fundamental in physics.   (Russell 1948: 27, 291)

How can mass really be nothing but concentrated energy that moves through space (Weyl 1922: 200), "just a very compact form of energy" (Thorne 1994: 172), "a form of bound energy" (Pagels 1982: 37), when mass rather than energy is the invariant quantity? (For that matter, unless a parcel of energy has a continuing identity through time, it cannot strictly be said to *move* at all, since motion requires that the bit of energy at one place and time be the selfsame thing as the bit of energy at another place and time.) Because energy is not Lorentz-invariant, it also seems difficult to endorse these remarks from Werner Heisenberg (a co-founder of quantum mechanics):

> Since mass and energy are, according to the theory of relativity, essentially the same concepts, we may say that all elementary particles consist of energy. This could be interpreted as defining energy as the primary substance of the world.   (1958: 67)[10]

Unfortunately, this kind of talk even finds its way into monographs for guiding physics teachers, one of which says, "Energy appears in still another wonderful disguise as mass itself" (Cotts and Detenbeck 1966: 66).

We should likewise not be satisfied with phrases like "the energy associated with the existence of the rest mass of a particle" (Feynman et al. 1963: I, p.15-11). "Associated" is a weasel word. In what sense are they "associated"? Is this a correlation between two different properties (just as bolts of lightning are associated with claps of thunder, but one is not identical to the other)? Or is one of these properties nothing but the other in disguise, or are they both reducible to some third property? Part of doing philosophy well is not allowing yourself to wriggle off the hook with vague phrases like "associated with."

In classical physics, a body's mass is often interpreted as the amount of some "stuff" of which the body is made. (Notice that classically, a body is made of *matter*, not of *mass*; its mass measures the amount of matter composing it.[11]) In relativity, however, this interpretation cannot be correct. That is because any property that represents the quantity of some "substance" must obey the following principle: The total quantity of some "stuff" in a system is the sum of the various quantities of that "stuff" belonging to the system's parts (where those parts are non-overlapping and together include the entire system).[12] A body's temperature, for example, does not measure the amount of some stuff it contains, since the temperature of my body is not equal to the temperature of its top half (98.6°F) plus the temperature of its bottom half (98.6°F), else I would be running an alarming fever (of 197.2°F). Likewise, a body's density does not measure the amount of some stuff that the body contains, since its density is not the sum of the densities of its parts. The same goes for velocity. On the other hand, in classical physics, mass is "additive" in this way, and so perhaps a body's mass can be interpreted as the amount of matter it is made of.

But in relativity, mass is not additive. To see this, let's work in an inertial frame in which $\boldsymbol{p} = 0$. (For a system of bodies, there always exists such a frame, and since mass is invariant, the result in that frame is applicable to any other inertial frame.) In that frame (as I have already mentioned), $\mathcal{E} = mc^2$, or

$$m = \mathcal{E}/c^2. \tag{8.3}$$

Consider a system of many constituents, where each exerts on the others only negligible forces when they are not in contact – such as a gas consisting of many molecules.[13] Since the system's total energy is the sum of the energies $\mathcal{E}_1, \mathcal{E}_2, \ldots$ of its constituents, equation (8.3) entails

$$m = (1/c^2)[\mathcal{E}_1 + \mathcal{E}_2 + \ldots]. \tag{8.4}$$

Recall that for any constituent (say, the ith one),

$$\mathcal{E}_i = m_i\gamma_ic^2 = m_ic^2/\sqrt{[1 - (v_i^2/c^2)]}. \tag{8.5}$$

Now for any x where $x^2 < 1$, we have the expansion

$$1/\sqrt{[1 - x]} = 1 + (\tfrac{1}{2})x + [(1 \cdot 3)/(2 \cdot 4)]x^2 + [(1 \cdot 3 \cdot 5)/(2 \cdot 4 \cdot 6)]x^3 + \ldots.$$

Let x be (v_i^2/c^2). If v_i is low compared to c, then the terms in the expansion with x^2 and higher powers of x are negligible. So equation (8.5) becomes

$$\mathcal{E}_i \approx m_ic^2 + (\tfrac{1}{2})m_iv_i^2. \tag{8.6}$$

Notice that this is just m_ic^2 plus the ith constituent's classical kinetic energy. Substituting for $\mathcal{E}_1, \mathcal{E}_2, \ldots$ in equation (8.4), we find

$$m \approx (1/c^2)[m_1c^2 + (\tfrac{1}{2})m_1v_1^2 + m_2c^2 + (\tfrac{1}{2})m_2v_2^2 + \ldots], \tag{8.7}$$

and so

$$m \approx [m_1 + m_2 + \ldots] + (1/c^2)[(\tfrac{1}{2})m_1v_1^2 + (\tfrac{1}{2})m_2v_2^2 + \ldots]. \tag{8.8}$$

Hence, the system's mass m exceeds the sum of its constituents' masses. (They differ by an amount that reflects the constituents' kinetic energies in the frame where the system has zero total momentum. This will shortly be important.) So in relativity theory, we cannot think of a body's mass as the amount of matter that forms it, where matter is a certain kind of "stuff" that is neither created nor destroyed.[14]

Bearing in mind that mass is not additive, consider the following argument, some version of which appears in most relativity texts. Suppose two bodies, each of mass m_0, crash into each other and stick together, forming one body of mass m_1 at rest. In the words of one fine text:

> [Although] the total energy $\mathcal{E}$ of the system of particles will ... be conserved [and] relativistic mass is conserved in this collision, *rest mass* [what I am calling simply "mass"] is not; m_1, the rest mass after the collision, is greater than $2m_0$, the rest mass before the collision. ... After the collision, no kinetic energy remains. In place of the "lost" kinetic energy, there appears internal (thermal) energy, recognizable by the rise in temperature of the colliding particles. ... [By $\mathcal{E} = mc^2$,] (increase in thermal energy) = (increase in rest mass)(c^2). Thus we see that the decrease in kinetic energy for this isolated system is balanced by a corresponding increase in mass energy.   (Resnick and Halliday 1985: 110)[15]

This text says that although the system is isolated, mass is not conserved! This is not correct. In this argument, the sum $2m_0$ of the masses before the collision is shown to be unequal to the mass m_1 of the body formed in the collision. But $2m_0$ is not "the rest mass before the collision" (that is, the mass of the system before the collision) because *mass is not additive*. The argument that mass is not conserved rests on the mistaken assumption of mass's additivity. To say that energy is conserved whereas mass is not helps to make energy seem like the real stuff of which bodies are made, leading to remarks like Heisenberg's (already quoted). Of course, it is also simply confusing for the text I have quoted to say on one page that the isolated system's rest mass is not conserved and on the very next page that "if we consider a closed system ... then we may regard the rest mass of the body (or of the system) as constant."

Since a body's mass should not be interpreted as its quantity of matter, relativity requires us to decouple mass from matter. This is not always done, as these examples show:

> According to the accepted mechanistic view, ordinary "ponderable" matter, i.e., the extended hard "stuff" with which we are all familiar,

constitutes the most fundamental layer of physical reality. . . . [M]atter
and energy were both conserved, but each separately from the other.
Einstein showed that the two levels could be regarded as identical. . . .
The stuff which appears to the senses as hard extended substance
and the quantity of energy which characterises a process are in fact one
and the same thing. Each of these two quantities can disappear and thereby give rise to an equivalent amount of the other.   (Zahar 1989: 261–2)

What relativity theory tells us is that matter is *yet another form of energy*,
mass energy. . . . Quantitatively, that is the meaning of the famous relation $\mathcal{E} = mc^2$.   (Rohrlich 1987: 73)

But the "famous relation" is between *mass* and energy, not between
matter and energy, since in relativity, mass is not the quantity of a
certain "stuff" or "substance": matter. (Nor is energy some sort of stuff,
since it is not Lorentz-invariant. And once again, the fact that mass
and energy can be interconverted does not demonstrate that one is
nothing but a form of the other.) The fact that when totaling a system's energy, we must include (along with the contributions of various
other forms of energy) a "mass energy" term mc^2 for each mass m in
the system (see equation 8.6) does not show that mass *is* energy any
more than it shows that energy *is* mass. To see which is an objectively
real property, we must note which is Lorentz-invariant.

If a body's mass is not the total quantity of some sort of "stuff" of
which it is made, then what is a body's mass? A body's mass is the
property it possesses that determines its "inertia" – that determines, in
other words, its resistance to being pushed around by a force. The
more massive the body, the more force is required to give it a certain
acceleration. In particular, relativity agrees with classical physics that
$F = p'$: the force on a body equals the rate at which its momentum
changes. (In relativity, however, $p = m\gamma v$, whereas in classical physics,
$p = mv$.) This conception of mass was held even by some classical
physicists. Maxwell, for example, said that mass is "the quantitative
aspect of matter" (1995: 811) but should not be understood as "the
quantity of matter in a body." Rather, "mass" is defined by its relations
to force, momentum, energy, and so on according to natural laws:

The Mass of a body is that factor by which we must multiply the
velocity to get the momentum of the body, and by which we must
multiply the half square of the velocity to get its energy.   (1995: 396; see
Harman 1985: 222–3)

When referring to a body's "mass," then, we must be thinking of that body as a thing that can feel a force and respond to it (by moving) *as a unit*.

Nevertheless, no macroscopic body is elementary; any macroscopic body is also a system of bodies. Its motion, then, is nothing but the motions of its constituents. A constituent's motion stands in a certain relation to its mass and the force on it. The remarkable fact is that this relation is *the same as* the relation of a macroscopic body's motion to its mass and the force it feels: $F = p' = (m\gamma v)'$. In other words, the law "scales up" (Feynman et al. 1963: I, p.19-2; Bohm 1996: 82–5, 110–18).

Let us see precisely why this fact is so remarkable. Macroscopic bodies are usually presumed to have elementary constituents (that is, constituents having no constituents themselves). In both classical and relativistic mechanics, there is a law relating the force felt by an elementary constituent, the constituent's motion in response to that force, and a *single additional parameter* characterizing the constituent. This need not have been. We can imagine a universe in which a constituent's motion in response to forces is affected in different ways by several of its properties, such as its shape, volume, and chemical activity. (Recall from chapter 3 that the laws of nature are not logically necessary.) In that case, there would be no single parameter capturing everything about the constituent that is needed for its motion to be related to the force acting on it. Moreover, even given that such a parameter exists for an elementary body, it does not follow that such a parameter exists for a macroscopic body. If a body's "mass" is defined as that single parameter, then even given the remarkable fact that the fundamental constituents of macroscopic bodies have masses, it is a remarkable fact that macroscopic bodies *have masses too*.

Furthermore, even given that macroscopic bodies' elementary constituents have masses *and* that macroscopic bodies have masses too, it does *not* follow that the law, by which an elementary constituent's motion is related to its mass and the force it feels, is *the same as* the law by which a macroscopic body's motion is related to its mass and the force it feels. If a body's "mass" is defined as the single parameter with which it is associated that plugs into whatever law relates an *elementary* body's motion to the force it feels, then once again, it is a remarkable fact that macroscopic bodies *have masses too*. Moreover, given that a macroscopic body has a mass, there is nothing inevitable about its mass being the sum of its elementary constituents' masses (as is the case classically but not relativistically).

It is certainly convenient for us that in order to predict a macroscopic body's motion in response to a force exerted upon it from outside (an "external force"), we do not need to determine the various forces exerted upon each of the body's constituents by the other constituents ("internal forces"). Rather, we can take the macroscopic body's total mass to be located at the macroscopic body's "center of mass," and then determine how this imaginary mass point would respond to the external force. Let us see why it is the case in classical physics that Newton's second law ($F = ma$, where F is a force on a body of mass m and a is the body's acceleration in response to that force) "scales up" in that it governs the motions not only of the elementary bodies, but also of the center of mass of a system of those bodies. To simplify the discussion, suppose that there are only three elementary bodies, each exerting forces on the others only when they collide. Let m_i be the ith body's mass, and let a_{ij} be the component of the ith body's acceleration that is caused by the force F_{ij} exerted by the jth body on the ith body. By Newton's second law, $F_{ij} = m_i a_{ij}$. By Newton's third law ("Every action has an equal and opposite reaction"), $F_{ij} = -F_{ji}$, and so $m_i a_{ij} = -m_j a_{ji}$. (In chapter 5, we saw that Newton's third law does not apply to electromagnetic interactions. But we have temporarily excluded these interactions by supposing that there is only action by contact between the bodies.)

Now consider bodies 2 and 3 as forming a single system of bodies. Let us see why that system's motion is likewise governed by Newton's second law. The system's center of mass is a kind of average of the positions of the system's constituents. Each constituent's contribution to the average is proportional to its mass. So the system's center of mass is a *weighted* average of its constituents' positions – weighted by their masses. If x_i is the position of the ith body, then the position x of the center of mass of the system composed of bodies 2 and 3 is given by

$$x = (m_2 x_2 + m_3 x_3)/(m_2 + m_3). \tag{8.9}$$

For Newton's second law to govern this system's motion would be for the force exerted on the system (namely, the sum of the forces exerted on the system's constituents by body 1, the body outside the system: $F_{21} + F_{31}$) to equal the system's mass ($m_2 + m_3$; in classical physics, mass is additive) multiplied by the acceleration a of its center of mass. From equation (8.9), it follows that

$$a = (m_2 a_2 + m_3 a_3)/(m_2 + m_3). \tag{8.10}$$

So Newton's second law governs the system's motion if and only if

$$F_{21} + F_{31} = (m_2 + m_3)a = (m_2 + m_3)(m_2 a_2 + m_3 a_3)/(m_2 + m_3)$$
$$= m_2 a_2 + m_3 a_3. \tag{8.11}$$

By Newton's second law applied to the constituents,

$$m_2 a_2 = F_{21} + F_{23}$$
$$m_3 a_3 = F_{31} + F_{32}. \tag{8.12}$$

So Newton's second law governs the system's motion if and only if

$$F_{21} + F_{31} = F_{21} + F_{23} + F_{31} + F_{32}. \tag{8.13}$$

But this is obviously true since (by Newton's third law) $F_{23} = -F_{32}$. So the system consisting of bodies 2 and 3 behaves in relation to body 1 in just the way that one elementary body behaves in relation to another. (By expanding the system to encompass more and more elementary bodies, we could work up to the conclusion that macroscopic bodies obey Newton's second law.) Thus, it takes the cooperation of Newton's third law for Newton's second law to scale up. Since either or both of these laws could have been different, there is nothing logically mandatory about the fact that the law relating a body's motion to the forces it feels scales up.

In relativity theory, a collection of bodies likewise behaves as a single body with the system's total mass m located at the collection's center of mass. A system's center of mass is again a weighted average of its constituents' positions – but relativity weights them by their energies (or, what comes to the same thing, their "relativistic masses"). So in place of (8.9), we have

$$x = (m_1 \gamma_1 x_1 + m_2 \gamma_2 x_2 + \ldots)/(m_1 \gamma_1 + m_2 \gamma_2 + \ldots). \tag{8.14}$$

For example, if the system is closed (that is, feels no external forces), then (since it behaves as a single body) its center of mass undergoes no acceleration. Notice that the weights in (8.14)'s weighted average are frame-dependent quantities (since the ith weight involves γ_i, which is a function of the ith body's velocity, which is obviously frame-dependent). So the location of the system's center of mass is not Lorentz-invariant. (In the next section, I will remove the restriction requiring the constituents to exert forces on one another only when they are in contact. That will enable us to discuss electromagnetic interactions and fields.)

The frame in which the system's total $\boldsymbol{p} = 0$ is the frame in which the system's center of mass is at rest. As we saw in (8.8), the system's mass increases as the kinetic energies of its constituents increase in the $\boldsymbol{p} = 0$ frame – that is, as its constituents move around more quickly relative to the system's center of mass. That is because in this frame, the system's mass is given by

$$m = \mathcal{E}/c^2 = (1/c^2)(\mathcal{E}_1 + \mathcal{E}_2 + \ldots), \tag{8.4}$$

so any increase $\Delta\mathcal{E}$ in the constituents' total kinetic energy contributes $\Delta\mathcal{E}/c^2$ to the system's mass, though it has no effect on any constituent's mass. Imagine, for example, a ball of hot gas, its molecules whizzing around randomly. We add to the constituents' total kinetic energy by heating the gas. We thereby increase the gas's mass. In what sense, if any, is this the "conversion" of energy into mass?

Suppose that, initially, we think of the gas as a *collection of bodies* by treating each gas molecule as having its own mass. In other words, we treat each of these bodies as individually feeling forces and being accelerated by them, its acceleration depending on its mass. Accordingly, we characterize the heat as having boosted various molecules' kinetic energies, but not their masses. Suppose we then change our perspective by considering the ball of gas as a *single body* that feels external forces, has a mass, and accordingly moves about as a unit. What was kinetic energy contributed by the heat in the original perspective becomes part of the gas's mass in the new perspective. This "conversion" of energy into mass is not the transformation of one kind of stuff into another, since neither energy nor mass is stuff of some kind or measures the amount of some kind of stuff (energy because it is not Lorentz-invariant, mass because it is not additive). But more importantly, we have just seen that this "conversion" of energy into mass is not a real physical process at all. *We* "converted" energy into mass simply by *changing our perspective* on the gas: shifting from initially treating it as many bodies to treating it as a single body.

Let's see that again. Suppose we begin by treating the gas as a single body. The body is heated. Heat energy flows into it and its mass increases by an equivalent amount. It looks like energy is being converted into mass; fluid, gossamer energy has "solidified" or "congealed" into matter, "the extended hard 'stuff' with which we are all familiar." But this is not a real process. Rather, it is an artifact of the perspective we have adopted. No such "conversion" occurs on a different perspective. Let us begin the gas's story again and this time, let us treat the gas as many bodies. We find no energy being transformed into matter as heat is being added to the gas – so long as we continue to regard the gas as many bodies. None of those bodies increases its mass while the gas is heating up. The heat energy goes into their kinetic energies relative to the gas's center of mass.

So, on the first perspective, energy was converted into mass, whereas on the second, no such conversion occurred. Furthermore, at whatever point in the story we choose to switch our perspective on the gas, we can make it appear that at that point, energy is being transformed into mass – even if at that point, heat is no longer being added to the gas! Let's begin by treating the gas as many bodies, and let's maintain this perspective throughout the heating of the gas. Suppose the heat source is then shut off. Nothing more is now happening to the gas. But at this point let us shift our perspective in telling the gas's story; henceforth let us treat the gas as a single body. In this shift, the kinetic energies (in the $p = 0$ frame) of the various gas molecules contribute an equivalent amount to (and so "become part of") the gas's mass. But obviously, no physical change accompanies this "transformation"; again, nothing is happening to the gas! This "conversion" of energy into mass is not a physical process. Thus, whether and when a "conversion" of energy into mass occurs in the story of the gas depends on the perspective from which we elect to tell that story and any shifts of perspective we make in the course of telling it.

The distinction between "internal" and "external" energy is often made by relativity texts (such as Resnick and Halliday 1985: 111–12). But the point is seldom made that because this distinction is a scientific convention, so also is the "conversion" of energy to mass or vice versa. As far as science is concerned, the line between bodies – between "internal" and "external" – is a convention: not built into the universe, but rather drawn onto it by us. The line between bodies is conventional because of the remarkable fact I have emphasized: that *any* system is characterized by a *mass* determining the way it responds (by moving as a unit – that is, by changing its velocity as a whole rather than its rotational speed or internal state) to the forces it feels as a whole (that is, to external forces). As Einstein says, "Every system can be looked upon as a material point as long as we consider no processes other than changes in its translation velocity as a whole" (1935: 225).

A shift in "perspective" (*not* reference frame!), in what are being treated as single bodies, obviously takes place in the course of the textbook discussion (quoted earlier) of two bodies colliding and subsequently sticking together. The text treated the system as initially consisting of two bodies, each with its own mass. But the text regarded the system after the collision as forming a single body. The "conversion" of energy into mass in this case is an artifact of this shift in perspective.

Sometimes, a shift in perspective is much subtler. Return to the tritium nucleus (one proton, two neutrons) that decays into a helium-3 nucleus (two protons, one neutron) and an electron and anti-neutrino that fly off at high speed. There is a "mass defect" in that the masses of a helium-3 nucleus, an electron, and an anti-neutrino add up to less than the mass of a tritium nucleus. The missing mass is said to have been "converted" into the kinetic energies of the resulting bodies. But this "conversion" of mass into energy is not real; it is an illusion produced by a subtle shift in our perspective. (The transformation of the tritium's neutron into a proton, an electron, and an anti-neutrino is, in contrast, a real occurrence – no illusion.) We treated the system as initially forming a single body: a tritium nucleus. But we treated the system after the decay as consisting of three bodies, each with its own mass. The system's mass after the decay is the same as the system's mass before the decay. There is no "mass defect" here; rather, mass is conserved. The "mass defect" appears to arise from the fact that the sum of the three masses after the decay is less than the system's mass before the decay (the difference reflecting the three bodies' kinetic energies in the $p = 0$ frame, according to equation 8.8).

But the sum of the three masses *after* the decay is also less than the system's mass *after* the decay. Mass is not additive, and it is our mistaken expectation that it is additive (arising because we expect it to measure the amount of stuff forming the bodies) that leads us to characterize the system as suffering from a "mass defect" – to ask where the "missing mass" has gone and to conclude that it has turned into energy. The "mass defect" results not from some physical transformation of matter-stuff into energy-stuff, but rather from our illicitly trying to view the system from two different "perspectives" at the same time. It is produced by our treating the post-decay system as a collection of bodies though we treated the pre-decay system as a single body. The fact that Δm of the system's initial mass "becomes" energy $(\Delta m)c^2$ when we think of the post-decay system as a collection of bodies, each with its own mass, does not mean that mass is really nothing but energy or that mass and energy are different ways of measuring the same property (like distance in feet and in miles). The "conversion" of mass into energy occurs because we have shifted our perspective, not because the nucleus has decayed.

A body's energy $m\gamma c^2$ is its kinetic energy plus a quantity of energy equal to mc^2 (equation 8.6). Since the latter quantity depends only on the body's mass, we can think of its mass as "associated with" a certain quantity of energy: a "mass energy" mc^2. Accordingly, it might sometimes help to think of relativity as having turned the classical law of energy conservation into the conservation of "mass-energy" by having added a term for the energy associated with a body's mass. In Einstein's words:

> Before the advent of relativity, physics recognised two conservation laws of fundamental importance, namely, the law of conservation of energy and the law of conservation of mass; these two fundamental laws appeared to be quite independent of each other. By means of the theory of relativity they have been united into one law.   (1931: 54)

How have they been "united"? In computing a system's mass for the purpose of applying the law of mass conservation, we cannot simply add the constituents' masses. But we can add their energies (in the $p = 0$ frame) and then divide by c^2 (equation 8.4). Hence, the law of energy conservation (in the $p = 0$ frame) entails and is entailed by the law of mass conservation. Likewise, in computing a system's energy for the purpose of applying the law of energy conservation, we must include not only terms standing for the more familiar forms of energy, but also terms of the form mc^2 (equation 8.6). Without these mc^2 terms, energy is not conserved.

But none of this means that mass really *is* energy (or "massergy") or "that mass is now viewed as a form of energy" (Spector 1972: 151). Mass is a real property whereas energy is not. Mass is no more a form of energy than it is a form of momentum. Energy and mass are not two sides of the same coin in the manner of E and B (as we found in chapter 7); a body's combination of energy and *momentum* in a given frame reflects its mass and that frame. Just as there is really only a single object, the electromagnetic field, which in different frames appears as different combinations of electric and magnetic fields, so there is really only a single property, the body's mass, which in different frames appears as different combinations of its energy and momentum. In this sense, relativity *unifies* energy, momentum, and mass. To think of the "conversion" of mass into energy (or vice versa) as a process that really occurs in nature, like the conversion of a caterpillar into a butterfly, would make sense only if energy and mass were (or measured the quantities of) real stuffs.

When an atomic bomb explodes, then, no mass is really converted into energy; the system after the explosion (what's left of the bomb and the city) has a mass equal to the mass of the system before the explosion (the unexploded bomb and the undamaged city). But if no mass is turned into energy, you might ask, where did the energy to cause the destruction come from? Energy, like mass, is a conserved quantity. The explosion redistributed energy within the system. In so doing, it turned some energy associated with masses of the system's constituents (mc^2) into other forms of energy, such as heat. (I have put this point picturesquely, as if energy were real and parcels of energy had continuing identities.) But this does not mean that the system's total mass diminished.

There is, however, one very important respect in which the foregoing account is incomplete.

5 Fields – At Last!

After the tritium nucleus decays, the sum of the resultant bodies' masses is *less* than the system's mass. Likewise, the sum of the gas molecules' masses is *less* than the mass of the ball of gas, and the sum of the masses of the two bodies prior to their collision is *less* than the mass of the single body formed in the collision. But consider another example of a "mass defect." A deuterium nucleus consists of one proton and one neutron. The sum of a proton's mass and a neutron's mass is slightly *greater* than a deuterium nucleus's mass. Our earlier result

$$m \approx (m_1 + m_2 + \dots) + (1/c^2)[(\tfrac{1}{2})m_1 v_1^2 + (\tfrac{1}{2})m_2 v_2^2 + \dots] \qquad (8.8)$$

does not leave room for $(m_1 + m_2 + \dots)$ to *exceed* m!

Here is another question that turns out to be related. We have seen relativity give a body's energy $m\gamma c^2$ as the sum of mc^2 and its kinetic energy (equation 8.6). Whatever happened to the body's potential energy?

Here is the answer. Take any case where the bodies in a closed system are exerting forces upon one another even though they are not in contact. In other words, take any case where the bodies possess various potential energies – or, perhaps it would be more correct to say, where the system as a whole possesses a certain potential energy. Then interpreting the system's potential energy as the field's energy $\mathcal{E}_f$ we find the system's total energy (the sum of the energies of its various constituents) to be $\mathcal{E}_f + m_1\gamma_1c^2 + m_2\gamma_2c^2 + \ldots$. In the $p = 0$ frame, the system's mass is given by

$$m = \mathcal{E}/c^2 = (1/c^2)(\mathcal{E}_f + m_1\gamma_1c^2 + m_2\gamma_2c^2 + \ldots). \qquad (8.15)$$

Let T_i stand for the ith body's kinetic energy (in this frame). Then (as we saw earlier in connection with 8.8), equation (8.15) entails

$$m = (m_1 + m_2 + \ldots) + (1/c^2)(T_1 + T_2 + \ldots) + (\mathcal{E}_f/c^2). \qquad (8.16)$$

In this light, reconsider the deuterium nucleus at rest and where T_1 (the proton's kinetic energy) and T_2 (the neutron's kinetic energy) are both zero. The fact that the nucleus's mass m is less than the proton's mass plus the neutron's mass entails (by 8.16) that $\mathcal{E}_f < 0$. The field we are talking about here corresponds to the force that holds the particles in an atomic nucleus together: the so-called "strong nuclear force." For our purposes, we can think of its field as like the electromagnetic field. Since the proton and neutron in the deuterium nucleus attract each other, their potential energy $\mathcal{E}_f$ is indeed a negative quantity. To see this, consider the analogous electromagnetic case: a system consisting of a stationary point body with positive charge q at a distance r from a stationary point body with negative charge $-q$. This system's electrostatic potential energy is $-q^2/r$ (as we saw in chapter 5), a negative quantity. So it makes sense that the deuterium nucleus's $\mathcal{E}_f < 0$. For any kind of nucleus, the quantity $mc^2 - [m_1 + m_2 + \ldots]c^2$ is sometimes called the nucleus's "binding energy." A stable nucleus has a negative binding energy: to break the nucleus apart, the nucleus must receive positive energy sufficient to cancel out its binding energy.

The field, then, joins the proton and neutron as a constituent of the deuterium nucleus. Apparently, the field has mass m_f given by Einstein's celebrated equation as $\mathcal{E}_f/c^2$. Since the field has mass, a real (Lorentz-invariant) property, the field is real. By equation (8.16), the nucleus's mass is the sum of the masses of its three constituents (in a frame where $p = 0$ and no constituent has kinetic energy). But hold on! Since the deuterium nucleus's mass is *less* than the proton's mass plus the neutron's mass, the field's mass m_f is *negative!* Sure, this quantity is Lorentz-invariant. But can the field be *real* by virtue of possessing *negative* mass?

Perhaps negative mass isn't so bad, considering that mass does not measure a quantity of stuff. Nevertheless, it turns out happily that we have made a mistake, which resulted in our attributing negative mass to the field. Since we are taking fields to be real (by virtue of their possessing mass), we must interpret the system's potential energy as belonging to the field rather than the bodies. Recall the analogous electromagnetic case: a positively charged point body (charge q) at a distance r from a negatively charged point body (charge $-q$), both at rest. If the electromagnetic field is real, we should interpret its energy as we learned to do in chapter 5: as $\int [1/8\pi][E^2 + B^2]dV$ (taking this integral over all of space), a positive quantity ($E^2 > 0$ everywhere; $B = 0$ everywhere since the charges are stationary). How can the field's energy also be $-q^2/r$, a negative quantity? Recall (from chapter 5) that a system's potential energy is the work (the force accumulated over a distance) needed to assemble the system, and so is defined only relative to the starting point (the "raw materials") for the assembly. To calculate the system's potential energy to be $-q^2/r$, we imagine the charges beginning very far apart, and we compute the work needed to bring them together arbitrarily slowly. Since they attract each other, the forces F exerted on the bodies in order to assemble the system in this manner must be *opposite* to the directions in which the bodies move during the assembly (the direction of ds). Hence, a bit of the work $F \cdot ds$ is a negative quantity, and so the potential energy (that is, the total work $\int F \cdot ds$ done in assembling the system) is negative. But this is only the potential energy *relative to* the starting point. When we think of the system as consisting of the two bodies *and* a field, the raw materials out of which we assemble the system are different. The starting point is the situation in which the field is everywhere zero, since in assembling the system, we must *make* the field. Therefore, the starting point (that is, the zero of potential energy) is when the two bodies are located at the *same* point, since then their charges cancel out and so $E = 0$ everywhere. Since the bodies attract each other, the forces F exerted on them in order to assemble the system (to separate them to a distance r) must be in the *same* direction as the bodies move during the assembly (the direction of ds). Hence, a bit of the work $\boldsymbol{F} \cdot \boldsymbol{ds}$ is a positive quantity, and so the total work needed to make the field equals the positive quantity $\int [E^2/8\pi]dV$.

From $(\mathcal{E}/c)^2 - p^2 = m^2c^2$ (equation 8.2), it follows that the electromagnetic field's total mass, energy, and momentum are related by

$$m_f = \sqrt{(\mathcal{E}_f^2/c^4 - p_f^2/c^2)}. \qquad (8.17)$$

In the frame where the two oppositely charged bodies are stationary, the magnetic field $\boldsymbol{B} = 0$ everywhere. Hence, the magnetic field's energy density $B^2/8\pi$ is zero everywhere, and so the only contribution to $\mathcal{E}_f$ is from the electric field. Also, that $\boldsymbol{B}$ is zero everywhere entails that the electromagnetic field's momentum density $\boldsymbol{G} = (1/4\pi c)\boldsymbol{E} \times \boldsymbol{B}$ (from chapter 5) is zero everywhere, and so the field's total momentum $\boldsymbol{p}_f = 0$. Therefore, equation (8.17) simplifies to

$$m_f = \mathcal{E}_f/c^2 = \int (E^2/8\pi c^2)dV, \qquad (8.18)$$

which is greater than zero because $E^2 > 0$ everywhere. Hence, the electromagnetic field's mass is a positive quantity. (Whew!)

So the electromagnetic field is a full-fledged constituent of a system, as real as the system's bodies since, like them, it possesses mass. Neither the field's energy density $(1/8\pi)(E^2 + B^2)$ nor its momentum density $\boldsymbol{G} = (1/4\pi c)\boldsymbol{E} \times \boldsymbol{B}$ is Lorentz-invariant, but its mass density is (see box 8.2). The electromagnetic field's total mass is less when the two oppositely charged point bodies are near each other (that is, when r is small) than when they are far apart, since the electric field's total energy is less when they are nearer (their charges more nearly canceling). (Don't be confused: the *force* exerted by each of the bodies on the other is *greater* when the bodies are nearer, even though the field's mass and

---

Box 8.2
Showing the electromagnetic field's mass density to be Lorentz-invariant

Begin with this version of equation (8.2) applied to the electromagnetic field's total mass, energy, and momentum:

$$m_f^2 c^4 = \mathcal{E}_f^2 - p_f^2 c^2.$$

This holds in any inertial frame. Now replace $\mathcal{E}_f$ with the field's energy density, p_f with the field's momentum density, and m_f with the field's mass density ρ_f:

$$\rho_f^2 c^4 = (1/64\pi^2)(E^2 + B^2)^2 - (1/4\pi c)^2(E \times B)^2 c^2. \tag{8.19}$$

Doing out the squares and recalling (from figure 2.6) that the magnitude of $E \times B$ is $EB \sin\theta$, where θ is the angle between the vectors E and B in the given frame, we find

$$\rho_f^2 c^4 = (1/64\pi^2)[E^4 + B^4 + 2E^2B^2] - (1/16\pi^2)E^2B^2\sin^2\theta. \tag{8.20}$$

As I mentioned in chapter 7 (section 2), $E \cdot B = EB \cos\theta$. So

$$(E \cdot B)^2 = E^2B^2 \cos^2\theta = E^2B^2 - E^2B^2 \sin^2\theta$$

(since for any angle, $\sin^2 + \cos^2 = 1$). So

$$(E \cdot B)^2 - E^2B^2 = -E^2B^2 \sin^2\theta. \tag{8.21}$$

Furthermore,

$$(E^2 - B^2)^2 = E^4 + B^4 - 2E^2B^2,$$

and so

$$(E^2 - B^2)^2 + 2E^2B^2 = E^4 + B^4. \tag{8.22}$$

Substituting (8.21) and (8.22) into (8.20) yields

$$\begin{aligned}\rho_f^2 c^4 &= (1/64\pi^2)\,[(E^2 - B^2)^2 + 4E^2B^2] + (1/16\pi^2)[(E \cdot B)^2 - E^2B^2] \\ &= (1/64\pi^2)\,(E^2 - B^2)^2 + (1/16\pi^2)(E \cdot B)^2. \tag{8.23}\end{aligned}$$

But $(E^2 - B^2)$ and $(E \cdot B)$ are the two Lorentz-invariant electromagnetic quantities I mentioned in chapter 7 (section 2). Hence, $\rho_f^2 c^4$ is Lorentz-invariant, and since c is Lorentz-invariant, ρ_f must be.

However, notorious difficulties arise for this demonstration when it is attempted for systems that include charges as well as fields, presumably because it neglects the forces holding a charged body together against the mutual electrostatic repulsion of its bits. See Romer (1995) and replies thereto.

energy are *less*.) To see precisely how this comes about, let us decompose the total E into $E_1 + E_2$, the components contributed by the two charged bodies. The field's total energy $\int [E^2/8\pi]dV = \int [(E_1 + E_2)^2/8\pi]dV = \int [E_1^2/8\pi]dV + \int [E_2^2/8\pi]dV + 2 \int [(E_1 \cdot E_2)/8\pi]dV$. Neither $\int [E_1^2/8\pi]dV$ nor $\int [E_2^2/8\pi]dV$ changes as the bodies move nearer, since each depends only on a single body. But $2 \int [(E_1 \cdot E_2)/8\pi]dV$ (the "cross term") *will* vary as the bodies move nearer. This quantity turns out (Zink 1967) to be $-q^2/r$, exactly the "binding energy": the negative quantity mentioned earlier as the system's potential energy relative to a zero level when the particles are infinitely far apart! This is a nice way to see the relation between the electric field's total energy (a positive quantity) and the binding energy (a negative quantity). As the bodies move nearer, r decreases, so q^2/r increases, so the field's total energy diminishes (but remains positive). The change in $\int [E^2/8\pi]dV$, as the distance between the two bodies changes, results exclusively from the change in the "cross term," and hence equals any change in $-q^2/r$. So energy *differences* (which are all that matter to energy conservation) are the same whichever way the field's energy is computed – as we saw in chapter 5.

Just as the electromagnetic field's mass is less when the two oppositely charged bodies are closer together, so the mass of the "strong nuclear force" field surrounding a proton and neutron is less when the proton and neutron are close together as a deuterium nucleus than when they are far apart. This helps to explain the deuterium nucleus's mass defect:

proton's mass + neutron's mass > deuterium nucleus's mass.

When this inequality refers to the "proton's mass" or the "neutron's mass," it means the particle *in isolation* from every other particle. That body (the particle-in-isolation) includes both the "bare" particle and the fields that it creates around itself. The same applies to references to the "deuterium nucleus's mass." (Likewise in equation (8.16): the constant terms $\int (E_1^2/8\pi)dV$ and $\int (E_2^2/8\pi)dV$ in the field's total energy are absorbed within "m_1" and "m_2," so that only the cross-term is reflected in "$\mathcal{E}_f$".) So since the mass of the deuterium nucleus's field is less than the mass of the isolated proton's field plus the mass of the isolated neutron's field, there is a "mass defect."[16] For example, if we approximate a stationary proton as a uniformly charged sphere of radius 7×10^{-14} cm, then its electromagnetic field's mass is $q^2/2Rc^2$, which comes to 1.7×10^{-29} grams, as compared to 1.67×10^{-24} grams for the proton's total mass. (Of course, the force holding a deuterium nucleus together is not electromagnetic, but rather the "strong" nuclear force.) That an isolated particle includes its field (in that portions of the particle's mass, energy, and momentum reside in the surrounding field) sounds something like Faraday's conception of a body as extending to infinity in virtue of its field (as I explained in chapter 6). Whether *all* of an isolated particle's mass is attributable to its various fields (electromagnetic, strong nuclear, and so forth) remains an open question.

In sum: When we think of the deuterium nucleus as a system of bodies rather than as a single body, we must ascribe masses not only to the proton and neutron, but also to the field. The masses of the proton and neutron in the deuterium nucleus are not their masses in isolation (which reflect not just the "bare" particles, but also their fields). A body's energy is mc^2 plus its kinetic energy, without a potential energy term, because that potential energy is ascribed to the field.

That a system consists not just of bodies, but also of fields, is reflected in the system's center of mass. As I noted earlier, the position x of a system's center of mass is the weighted average of its constituents' positions x_i – weighted, in relativity theory, by their energies:

$$x = (m_1\gamma_1 x_1 + m_2\gamma_2 x_2 + \ldots)/(m_1\gamma_1 + m_2\gamma_2 + \ldots). \qquad (8.14)$$

But this formula applies only when there are no fields by which the constituents can exert forces on one another without being in contact. When the system includes a field – for example, the electromagnetic field – then the distribution of the field's energy affects the location of the system's center of mass (Einstein 1989: 206):

$$x = \frac{m_1\gamma_1 x_1 + m_2\gamma_2 x_2 + \ldots + \int x[E(x)^2 + B(x)^2](1/8\pi)dV}{m_1\gamma_1 + m_2\gamma_2 + \ldots + \int [E(x)^2 + B(x)^2](1/8\pi)dV} \qquad (8.24)$$

In other words, the system's center of mass is the average location of the system's energy. Some of that energy belongs to masses: the ith particle's $m_i c^2$ and kinetic energy are at x_i, the ith particle's location. The rest of the system's energy belongs to the field: at a given point x, the electromagnetic field's energy density is $[E(x)^2 + B(x)^2]/8\pi$, where $E(x)$ and $B(x)$ are the electric and magnetic field strengths at x. (These field strengths are not Lorentz-invariant, but neither are the γ_is and neither is the center of mass's position.)

A field is not *made* of energy any more than it is made of momentum. Since matter is not some sort of stuff, a field is not *made of* matter either. But if matter is defined as anything with mass, then a field is matter.[17] Since the electromagnetic field has mass, the field is ontologically on a par with bodies; since the electromagnetic forces felt by bodies are caused by the local electromagnetic field, electromagnetic interactions obey spatiotemporal locality according to relativity theory!

Let us pause momentarily to savor this result. In chapter 5, we tried to use the ontological status of energy and momentum to support the field's reality (see box 8.3). But this argument ultimately failed when we learned from relativity theory that energy and momentum are frame-dependent, and hence unreal. Relativity did, however, reveal mass to be real. Its ontological status has now come to underwrite the electromagnetic field's reality, since the field has turned out to possess mass. Spatiotemporal locality is thus (at last!) secured for electromagnetic interactions.

Box 8.3
A loose end

At the close of chapter 5, we considered using the remarkable relation $S = G\,c^2$ to remove the ambiguity in the formula for the field's energy flux density (that is, the amount of field energy per second that is crossing a surface, per unit of surface area). Having found $G = (1/4\pi c)E \times B$ to be the field's momentum density, we argued that the field's energy flux density is given by the Poynting vector $S = (c/4\pi)E \times B$ rather than by S plus some non-zero divergence-free F. Although any $S + F$ is a solution to the continuity equation for energy, S is the only solution that makes the field's energy flux density equal to the field's momentum density times c^2. Though classical electromagnetic theory could not explain it, this relation between energy flux and momentum seemed altogether too neat for us to allow it to be disrupted by a non-zero F.

This intuition is borne out by relativity theory. As we saw at the start of section 3, energy and momentum are as unrelated in classical physics as space and time, but are unified by relativity theory in the manner of space and time. One consequence of this unity is that the energy flux density is the momentum density times c^2.

Picture "relativistic mass" $m\gamma$ (which, of course, is not an invariant quantity) as flowing with constant density ρ and uniform velocity v down a pipe. How much "relativistic mass" flows out of the pipe in time interval Δt? One way to calculate this quantity is as the "relativistic mass" flux density (the rate at which "relativistic mass" is flowing out across each unit area of a surface stretched over the pipe's exit) times the area A of the surface over which the "relativistic mass" is flowing times Δt – that is, as "relativistic mass" flux density times $A \Delta t$. Another way to calculate how much "relativistic mass" flows out of the pipe in Δt is as the quantity of "relativistic mass" in the portion of the pipe extending from the exit back up the pipe to a distance $v \Delta t$ – that is, $\rho v A \Delta t$. Since these two methods of calculation must agree,

$$\text{"relativistic mass" flux density} = \rho \mathbf{v}. \tag{8.25}$$

But relativity says

$$\text{momentum} = m \gamma \mathbf{v}, \tag{8.26}$$

and

$$\text{"relativistic mass"} = m \gamma. \tag{8.27}$$

So

$$\text{momentum} = \text{"relativistic mass" } \mathbf{v}. \tag{8.28}$$

Letting $\mathbf{G}$ stand for the momentum density and ρ for the density of "relativistic mass," (8.28) entails

$$\mathbf{G} = \rho \mathbf{v}. \tag{8.29}$$

From (8.25) and (8.29),

$$\mathbf{G} = \text{"relativistic mass" flux density}. \tag{8.30}$$

Now "relativistic mass" $= m \gamma = \mathcal{E}/c^2$, and so

$$\text{"relativistic mass" flux density} = \text{energy } (\mathcal{E}) \text{ flux density}/c^2. \tag{8.31}$$

From (8.30) and (8.31),

$$\mathbf{G} c^2 = \text{energy flux density}, \tag{8.32}$$

just as we suspected at the close of chapter 5.

6   Erasing the Line between Scientific Theory and its Philosophical Interpretation

One final point. The electromagnetic field, in virtue of its mass, possesses a gravitational influence (according to the general theory of relativity), and so affects the motion even of *uncharged* bodies (Feynman et al. 1963: II, p.27-6). So an exquisitely sensitive experiment could tell us the mass distribution in a region containing an electromagnetic field. It would also tell us whether it is correct to assign the zero level of potential energy ("sea level") to the condition where the field is zero, since the absolute *amount* of mass, not mere mass differences, would affect the gravitational force felt by a test body.

This is a nice example of why it is problematic to say that two interpretations of a given theory are "empirically equivalent" – that is, make all of the same observable predictions. Although two different assignments of energy's zero level yield the same observable predictions in classical physics (where only energy differences matter), they lead to different observable predictions when combined with relativity theory. No two interpretations of a given theory are empirically equivalent in connection with *all possible* hypotheses that might be added to them. (However, two interpretations may be empirically equivalent relative to a certain limited set of additional hypotheses.)

At the end of chapter 5, we noticed that by way of the hypothesis that (field energy flux density) = (field momentum density)c^2, we could regard observational evidence that G is the field momentum density as confirming that there exists a circulation of energy around a magnet near a stationary charge, even though that energy circulation would not be detectable directly. We drew the general moral that no hypothesis can ever be known *for sure* to be beyond the reach of being tested against observations by virtue of concerning matters (such as an "idle" energy circulation) far removed from what we can observe directly. Given any hypothesis, no matter how remote its subject matter, we cannot conclusively rule out the possibility that eventually, theories will be confirmed that enable that hypothesis to make an observable difference, allowing predictions regarding our observations to be coaxed out of it. With the example of the electromagnetic field's gravitational influence allowing energy's absolute level to make a (minute) difference to the motions of bodies, we have now seen that this point regarding the testability of a scientific hypothesis applies equally well to a philosophical interpretation of some scientific theory. Two interpretations of the same theory may eventually be made to yield different observable predictions – by being combined with some new, well-confirmed theory.

At the outset of this book, I described a principal task of the philosophy of physics as the *interpretation* of physical theories. It might then have seemed to you that two interpretations cannot apply to *the same theory* if they make *different* observable predictions; rather, they must amount to *different physical theories*. But we cannot ever know for certain that two interpretations will always continue to yield exactly the same predictions regarding our observations, even as additional theories become well confirmed. The sharp, permanent line between a scientific theory (whose business it is to predict our observations) and its philosophical interpretation (which specifies what reality would or could be like if the theory succeeds in predicting our observations) breaks down if new empirical predictions can be derived from a theory's interpretation. The interpretation of a physical theory is just more physical theory!

The interpretation of physical theories can thus lead to innovation in physics – as we also saw at the end of chapter 7, where we witnessed relativity theory arise from Einstein's dissatisfaction with "asymmetries" in the customary interpretation of classical electromagnetic theory. The interpretation of scientific theories, although traditionally "philosophical" in its concern with rigorous conceptual foundations, is also an essential component of doing science.

Discussion Questions

You might think about . . .

1   If energy is unreal (since it is not Lorentz-invariant), why do we have to pay for the energy (measured in kilowatt-hours) that flows into our homes to be converted there into other forms by our appliances? What are we paying for?

2   Consider the property of moving at 5 m/s relative to a given inertial reference frame – call it frame F. (We might just as well have considered the property of having 5 joules of kinetic energy relative to frame F.) This property is Lorentz-invariant *in a cheap sense*, since in any inertial frame (whether F or any other), the body is moving at

5 m/s *relative to F*. By virtue of its Lorentz invariance, does a body's speed relative to frame F qualify as one of its objectively real properties? Could this property be causally relevant? Is it an *intrinsic* property of the body?

3   Consider the situation described in chapter 5, question 6. Show that in different inertial frames, the path of energy flow is different (Taylor and Wheeler 1966: 147).

4   Contrast these ideas: (a) In a nuclear reaction, the energy associated with mass is converted into other forms of energy. (b) In a nuclear reaction, mass is converted into energy. (c) In a nuclear reaction, matter is converted into energy. Tease apart the various concepts at play in this passage from a standard physics textbook:

> This equation [You know which one!] tells us that mass is a form of energy and that a particle of mass m has associated with it a rest energy $\mathcal{E}_0$ given by mc^2. This rest energy can be regarded as the internal energy of a body at rest. Thus the electron and positron have internal energies merely because of their masses. . . . [This equation] asserts that energy has mass. We therefore conclude that the conservation of energy is equivalent to the conservation of mass.   (Halliday et al. 1992: I, 166–7)

5   Consider the collision of two bodies. They may be billiard balls or elementary particles, so long as they are larger than points. Suppose that the bodies are rigid. (In other words, they do not deform, compress, etc., when they collide; they retain their shapes exactly.) Then a force is produced on body A by its collision with body B. Since body A is rigid, it instantly accelerates in response to this force.

(a)   Consider this argument: The force on body A occurs where it touches B. But since A is rigid, one effect of this force occurs at the edge of A that is farthest from where A touches B: this part of the body accelerates along with the rest. So we have a separation in space between a cause and its (instantaneous) effect. Does this violate spatial locality?

(b)   Since the bodies are rigid, their collision is supposed to be simultaneous with the acceleration of their edges located farthest from the collision. Does relativity theory permit this? What is the upshot?

6 The kinetic energy of an oxygen molecule in a ball of gas (in the frame where the ball's $p = 0$) contributes to the ball's mass. So why doesn't the kinetic energy of an oxygen molecule moving outside any ball of gas contribute to the molecule's mass?

Notes

1 Einstein's theory is relevant to many important issues in the philosophy of physics. I shall touch on only a few of them, and I shall confine myself almost entirely to Einstein's 1905 "special" theory of relativity, leaving aside the philosophical ramifications of his 1915 "general" theory. For rich discussions of many related issues (from which I have also occasionally borrowed in the first two sections of this chapter), see Earman (1989), Geroch (1978), Maudlin (1994), and Sklar (1976). For an elementary philosophical introduction to special relativity, see Salmon (1980).

2 The relativity of motion also entails that E and B are not separate real entities, as we discussed in chapter 7. But Heaviside did not anticipate Einstein's "asymmetry" argument. In contrast to Heaviside's view of energy, Maxwell recognized that only energy *differences* have observable consequences and yet apparently did not hesitate to believe that there was a hidden fact regarding the absolute quantity of energy possessed by a system:

> We cannot reduce the system to a state in which it has no energy, and any energy which is never removed from the system must remain unperceived by us, for it is only as it enters or leaves the system that we can take any account of it. We must, therefore, regard the energy of a material system as a quantity of which we may ascertain the increase or diminution as the system passes from one definite condition to another. The absolute value of the energy in the standard condition is unknown to us, and it would be of no value to us if we did know it, as all phenomena depend on the variations of the energy, and not on its absolute value. (1892: 167–8, also 134, 166)

3 My thanks to an anonymous referee for suggesting this example.

4 The distance d between the mirrors is the same in both frames, since it is at right angles to the direction in which one inertial frame is moving relative to the other.

5 This omits potential energy, which is to say, the field's energy – the subject of section 5 of this chapter.

6   I have been using "system" and "body" interchangeably in this paragraph. I shall say more about this in the next section.

7   A related point is that p's conservation in every inertial frame entails $\mathcal{E}$'s, since p in one frame is a function of p and $\mathcal{E}$ in another (Taylor and Wheeler 1966: 113–15). This entailment leads Einstein to say, "The principles of the conservation of momentum and of the conservation of energy are fused into one single principle" (1949: 61).

8   Here's a related passage:

> Does [$\mathcal{E} = mc^2$] mean that mass and energy are the same thing? . . . [I]n order for this relation to imply a complete identity, it seems that a total conversion of mass into energy should be possible. However, such a conversion is not possible, because of the conservation of baryon number, which says that the total number of these particles cannot change; they can be converted into other baryons, but cannot disappear entirely. Baryons include, among other things, protons and neutrons, which make up most of the mass of ordinary matter, and therefore most of the mass of ordinary matter is not available for conversion into energy. . . . So we are left in the rather odd position of asserting an equivalence between mass and energy, but not an identity that would allow one to be fully transformed into the other.   (Morrison 2000: 182)

This seems incorrect on two counts. (i) Baryon number conservation does not prevent baryons (such as protons and neutrons) from being "fully transformed" into energy (or vice versa). A proton has baryon number +1 and an anti-proton has baryon number −1, so their mutual annihilation (with the creation of light, with baryon number 0) does not violate baryon number conservation. Of course, anti-protons are not "ordinary matter," and so the passage is correct insofar as it is suggesting that without the assistance of extraordinary matter, most of the mass of ordinary matter is not available for conversion into energy. (ii) That mass and energy are the same thing does not require that a total conversion of mass into energy be possible. Even if mass and energy are essentially the same (somewhat like water and steam), it could be a natural law that 90 percent is the maximum fraction of a quantity of mass that a process can turn into energy. Or there could have happened not to be a process that turns mass entirely into energy, just as there could happen not to be a process that turns diamond into graphite (though both are forms of carbon).

9   Of course, there is an objective fact regarding energy content *in the $p = 0$ frame*, but this is a cheap sort of objectivity: it holds in all frames because it refers to a specific frame. (Recall box 8.1.) Notice also that relativity's definition of energy as $m\gamma c^2$ could, as far as energy conservation is

concerned, just as well have been $m\gamma c^2 + K$ for any constant K; we saw in chapter 5 that any such constant cancels out. But if we stipulate that when $m = 0$, energy becomes zero, it follows that $K = 0$ (Einstein 1935: 225, 229).

10   The difficulties for understanding energy as a locally well-defined, conserved quantity become even more substantial (!) as we pass to the *general* theory of relativity (Hoefer 2000).

11   Here is Definition 1 in Newton's *Principia*, the revolutionary work in which he proposed his laws of motion and gravity: "*Quantity of matter is a measure of matter that arises from its density and volume jointly.* . . . I mean this quantity whenever I use the term 'body' or 'mass' in the following pages" (1999: 403–4; italics in original).

12   See Campbell (1957: 282–3, 286), Meyerson (1962: 179), and Cartwright (1975: 147–9). That a property is "additive" is necessary but insufficient for it to measure the quantity of some stuff. Energy, for instance, is additive but does not measure the quantity of some stuff because energy is not Lorentz-invariant. Likewise for "relativistic mass."

13   This restriction means that there is no potential energy, there are no fields, and so forth; in classical terms, the only energy is kinetic energy. Of course, in the next section, we will have to talk about fields and potential energy, since as we have seen in previous chapters, they are crucial to whether spatiotemporal locality holds.

14   Contrary to Mackie (1973: 152).

15   I have substituted "m_1" for their symbol for the resultant body's mass. A similar argument appears in Eyges (1972: 242).

16   Suppose we begin with a proton and a neutron far apart at rest. The system's mass is the sum of the "bare" particles' masses and the field's mass. Given a negligible nudge, the particles move nearer and are bound together. The field's mass is now less than before. The system's mass is the sum of the "bare" particles' masses and the field's mass (*when the particles have no kinetic energy*). So since the field's mass has diminished, hasn't mass conservation been violated? No. After fusion occurs, the mass of the system is not equal to the sum of the proton's mass, the neutron's mass, and the field's mass, since each body now has kinetic energy.

17   Einstein writes: "[W]e denote everything but the gravitational field as 'matter.' Our use of the word therefore includes not only matter in the ordinary sense, but the electromagnetic field as well" (1952: 143).

9
Quantum Metaphysics

Electromagnetism was the nineteenth century's challenge to spatiotemporal locality. Early in the twentieth century, relativity theory supplied a powerful argument that electromagnetic interactions obey locality. But quantum mechanics also arrived on the scene. It posed an entirely new challenge to locality. As the twenty-first century opens, natural philosophers continue to wrestle with this challenge.

Quantum mechanics (QM) was developed in the 1920s, principally by Werner Heisenberg (German physicist, 1901–76), Erwin Schroedinger (Austrian physicist, 1887–1961), and Max Born (German physicist, 1882–1970). It gave the first successful explanations of the chemical bond, the Periodic Table of the Elements, and the spectrum of hydrogen. It has since been used to understand phenomena ranging from semiconductors and superconductors to white dwarf stars and radioactive nuclei. Its postwar offspring, quantum electrodynamics, is in staggeringly precise agreement with experimental outcomes and is the foundation of all current theories of elementary particle interactions.

Notoriously, interpretations of QM portray the universe as a very weird place. We shall focus, of course, on whether QM reveals that spatiotemporal locality is violated. Research directed at this question has led to "experimental metaphysics."

1 Is Quantum Mechanics Complete?

One of science's goals is to predict what we will see, hear, and so forth on various occasions. Some things we observe directly: we use only our unaided senses. Sometimes we look directly only at a measuring device (such as a galvanometer), but we think of ourselves as thereby measuring something about a certain "system" (such as a current-carrying wire) that interacts with the device. The "outcome" of the measurement is the state of the measuring device after it has interacted with the system, a state that we can ascertain by observing the device directly. Ideally, the measurement outcome is perfectly correlated with the state of the system when it interacted with the device. (For example, the galvanometer needle points to the numeral "5" printed on the galvanometer's dial if and only if the wire carries 5 amps of current.) The measurement outcome is the directly observable thing that QM aims to predict.

QM makes these predictions in terms of *likelihoods*. Take a device that we would ordinarily interpret as measuring a system's current, or position, or whatever – one of the system's "observable quantities" (observable indirectly, that is, by way of measuring devices). Let the device have two light bulbs mounted atop it: suppose we would ordinarily interpret the *left* bulb's lighting as the measurement outcome indicating that the observable quantity's value falls *inside* a certain range (say, 5 to 10 amps inclusive) and the *right* bulb's lighting as indicating the value to be *outside* that range. QM might predict only that the left bulb's lighting is 70 percent likely – not that it is certain (100 percent likely) or certain not to happen (0 percent likely).

Usually, we take it for granted that at each moment, the system possesses some *definite* quantity of the observable being measured (current, position, etc.). Common sense also suggests that the definite quantity possessed at a moment arbitrarily near to when the measurement was made was arbitrarily near to what the measurement outcome indicated (say, 5 to 10 amps). The device merely revealed a property that the system already possessed even before it was measured to have it, a property that the system would still have possessed at the time of measurement even if it had not then been measured at all. But these ideas about the relation between reality and what QM directly describes (the likelihoods of various measurement outcomes) are ideas about the *interpretation* of QM. We cannot take them for granted here.

For example, suppose the system consists of a single point-sized particle, and suppose the observable is its position. Ordinarily, we would take it for granted that at any moment, either the particle is inside a certain region of space (which includes its borderline) or the particle is outside it. QM would then be *incomplete* in that it cannot be interpreted as telling us *definitely* whether the particle is inside or outside. QM gives us only *likelihoods*: the probability of the particle's being inside (or outside) the region – or, I should rather have said, the probabilities of getting various measurement outcomes with a device that we ordinarily interpret as measuring the particle's position.

However, there is another, more outlandish possible interpretation: that when QM predicts that the "inside" measurement outcome is 70 percent likely, QM supplies *complete information* regarding the particle's location just before the measurement interaction occurs (as far as the particle's being inside or outside the region is concerned). How could that be? Because the particle has *no* definite location then. I don't mean that part of the particle (70 percent of it) is inside the region and the rest outside, nor do I mean that the particle is moving very quickly and 70 percent of the time, it is inside the region. Rather, I mean that *the only fact there is* about its location (as far as being inside or outside the region is concerned) is the fact that QM tells us: before the measurement, it is in a state such that were its location measured, there is a 70 percent chance that the measurement's outcome would be the "inside" bulb's being lit. Only *after* the measurement is made does the particle possess a definite location (as far as being inside or outside is concerned): if the "inside" bulb lights, then just after the measurement, the particle is definitely inside the region.

On this second interpretation, QM supplies *complete information* regarding the particle's pre-measurement state. That weird state, involving no definite position, is called a "superposition." In QM symbols, the superposition might be expressed thus:

$$(\sqrt{0.7})|\text{inside}> + (\sqrt{0.3})|\text{outside}>.$$

This notation expresses the particle's pre-measurement "wave-function" (also known as its "ψ-function" or "state-function"). On this interpretation, the particle's wave-function at some moment *completely* describes the particle's state at that moment. (For simplicity, I have given only the "part" of the wave-function concerned with the particle's location inside or outside the region.) The "|inside>" and "|outside>" are the wave-functions expressing the "pure" states of *definitely* being inside and outside the region, respectively; measuring a particle whose wave-function is simply |inside> is *certain* to result in the "inside" bulb's being lit. The squares of $(\sqrt{0.7})$ and $(\sqrt{0.3})$, namely 70 percent and 30 percent respectively, are the probabilities that an ideal measuring device would have its "inside" or "outside" bulb lit, respectively, after interacting with a particle in the superposition. Since $\sqrt{1} = 1$, the pure state |inside> is equal to $(\sqrt{1})$|inside>, and so the outcome of measuring a particle in this state is $(\sqrt{1})^2 = 100$ percent likely to be the "inside" bulb being lit.

If the measurement outcome is the "inside" bulb being lit, then the particle's wave-function becomes |inside> after the measurement. The particle is then definitely located inside the region. But if QM is complete, the particle was not inside *before* the measurement; its state was the superposition. It seems, then, that the measurement interaction *causes* the particle's state to change from the superposition to the pure state. This apparent change is known as the "collapse of the wave-function." On the other hand, if QM is incomplete so that the particle was definitely inside even before the measurement was made, then the collapse of the wave-function is not a physical change in the particle. Rather, it is merely a change in our *knowledge* of the particle's state: before the measurement, we are 70 percent confident that the particle is inside the region, but after the measurement, our confidence rises to 100 percent certainty.

If QM is complete, then when the particle in a superposition interacts with the measuring device, the particle's state prior to the interaction (together with the device's prior state) fails to *determine* the measurement outcome. The particle's wave-function will certainly collapse, but there is an ineliminable element of chance in what pure state it will collapse into. For example, suppose we have two particles, *A* and *B*, that at a given moment are in exactly the same position state according to QM: each particle, if measured, is 70 percent likely to give the "inside" outcome. Suppose we measure both particles and the outcome is *A* "inside," *B* "outside." If QM is complete, then prior to their measurement, there was no difference between the two particles to explain the difference in their subsequent measurement outcomes. There *is* an explanation of why *A* gave the "inside" outcome: it was in a state in which it was 70 percent likely to do so. Likewise, there is an explanation of why *B* gave the "outside" outcome: it was in a state in which it was 30 percent likely to do so. (Unlikely events *do* sometimes happen; they are just unlikely.) But if QM is complete, then before the measurement, there was no difference between the two particles in some respect that QM leaves out – some "hidden variable" – that could explain why *A*'s outcome differed from *B*'s. Ordinarily, we expect there to be such explanations. When Jones develops lung cancer and Smith does not, we expect there to be a reason for the difference: say, that Jones was a heavy smoker and Smith never smoked. But QM governs a carcinogen's interaction with DNA, resulting ultimately in cancer. So if QM is complete, there may be no difference whatsoever between Jones and Smith to account for their different outcomes. In principle, Jones and Smith could have been *exactly alike* in all of their risk factors, genetic and environmental – even in those risk factors that we have not yet discovered. Jones was simply unlucky.

If QM is complete, then a superposition is a weird physical state. What does it mean to say that a particle is neither simply inside the region nor simply outside it, but instead is in a state such that it is 70 percent likely to produce the "inside" measurement outcome (and then be definitely inside)? We can sharpen the question by exploiting a famous thought experiment suggested by Schroedinger (Wheeler and Zurek 1983: 152–67): the "paradox of Schroedinger's cat." Suppose we have a radioactive atom with a half-life of one hour. In other words, it is 50 percent likely to decay during that period. A cat inside a cage is hooked up to a device that will expose it to lethal gas when and only when the atom decays. If the atom, device, and cat are left unobserved, then after one hour, the entire apparatus is in a superposition: were we to examine the apparatus, there is a 50 percent chance that the measurement outcome will be that the atom has decayed and the cat has been killed, and a 50 percent chance that the outcome will be that the atom has not decayed and the cat has not been killed. But what in the world can it mean for a cat to be in a superposition of being alive and being dead?

It might be objected that the cat is an observer, continuously observing herself and the device, and so the cat's wave-function is always in a pure state with regard to life and death. This raises a further delicate issue, the "measurement problem": what makes something a measuring device? When two particles interact, there is no wave-function collapse. But when a particle interacts with a measuring device, the measurement has a definite outcome: the particle's wave-function collapses into a pure state of whatever the device measures. But how can this be, if a measuring device is nothing but a bunch of particles?

In other words, suppose a subatomic particle interacts with a macroscopic object, such as a cat or a measuring device. (We might even think of the cat as a "device" for measuring whether or not the atom has decayed.) If we predict the result by applying the standard QM machinery (the so-called 'Schroedinger equation"), then the predicted result is a superposition, such as a 50–50 superposition of |atom decayed>|cat dead> and |atom not decayed>|cat alive>. But when we look, we obviously see a cat that is in some definite state (alive or dead), not in some superposition (whatever that would look like). So unless we say that a measurement is accompanied by wave-function collapse, QM will conflict with the outcome of every measurement that is ever made! On the other hand, if we say that a measurement does not proceed according to the Schroedinger equation but instead is accompanied by wave-function collapse, then what counts as a measurement? Is interaction with a consciousness required, or will any macroscopic object do? If the former, then consciousness must play some special role in the laws of nature. If the latter, then there must be a sharp line between the microscopic and the macroscopic. The measurement problem is the problem of interpreting or otherwise dealing with the prospect of superpositions involving microscopic and macroscopic entities coupled together – their states "entangled" (in Schrodinger's happy phrase).

In short, the proposal that QM is complete raises several difficult questions. Yet this bold view was put forward by Born and Heisenberg (1928: 178) and was famously elaborated by Niels Bohr (Danish physicist, 1885–1962) in what became known as the "Copenhagen interpretation" of QM. It became widely accepted among physicists; in the words of Sir Rudolf Peierls (1907–95), who studied QM from Heisenberg and Bohr:

> I object to the term Copenhagen interpretation [because] this sounds as if there were several interpretations of quantum mechanics. There is only one. There is only one way in which you can understand quantum mechanics. There are a number of people who are unhappy about this, and are trying to find something else. But nobody has found anything else which is consistent yet, so when you refer to the Copenhagen interpretation of the mechanics what you really mean is quantum mechanics. And therefore the majority of physicists don't use the term; it's mostly used by philosophers.   (Davies and Brown 1986: 71)

The most notable person who was "unhappy about this" was Einstein. He said that for QM to be complete, there would have to be "spooky actions at a distance" (Born 1971: 158). It might surprise you that Einstein, who arrived at relativity by eliminating "surplus structure" (as we saw in chapter 7), then wished to add "surplus structure" (in the form of hidden variables) to QM. But Einstein had a reason for regarding this structure not as surplus, but as indispensable: it is needed to avoid an especially spooky violation of spatiotemporal locality.

To see Einstein's point, let us look at another observable quantity. When an electron (or similar particle) enters a certain device equipped with a non-uniform magnetic field (a "Stern–Gerlach apparatus"), the electron ends up either exiting the top of the device, triggering the particle detector there, or exiting the bottom of the device, triggering the detector there. Let each detector have a bulb that lights when that detector is triggered. Ordinarily, which bulb lights is interpreted as indicating the electron's "spin" (a property somewhat unlike the familiar property of the same name).[1] Electrons triggering the top detector are said to have a (+) spin component in the direction in which the device's magnetic field is set – let's say, the x axis – while electrons reaching the bottom detector have a (−) spin component along that axis. The device – a "spin detector" – could have its magnetic field reoriented to measure a different spin component: say, along the y or z axis. But whatever the device's setting, electrons arrive either at the top detector or the bottom one; there are only two values that an electron's spin component can be measured to have. Common sense might lead us to expect a given electron at any moment to have some definite combination of spin components – say, (+) for x-spin, (−) for y-spin, and (−) for z-spin. Of course, QM predicts only various probabilities – such as the probability that the device, with its magnetic field oriented to measure x-spin, will have its (+) bulb turn on. QM is incomplete if a given electron possesses a definite value of x-spin but QM specifies only some probability (other than 0 percent and 100 percent) of an x-spin measurement's returning (+).

We want to understand Einstein's argument that QM's completeness would require "spooky action at a distance." So for the sake of this argument, let us interpret QM as complete: a particle's spin state is completely captured by the particle's wave-function given by QM. Its spin components (such as its x-spin) need not be in pure states (such as $|x+ >$).

Consider a source emitting *pairs* of particles: one particle of each pair goes toward the left, the other toward the right (figure 9.1). Regarding certain specially prepared sources, QM says that a measurement of each particle's x-spin is 50 percent likely to return the (+) outcome and 50 percent likely to return (−), and the same probabilities apply to its y-spin and z-spin. So if QM is complete, particles in a given pair are not in pure states for any spin component prior to undergoing a spin measurement. If a particle's x-spin (for example) is

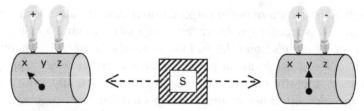

Figure 9.1 A source S emits pairs of particles toward the detectors on the two wings of the apparatus. The detectors can be set to measure x-spin, y-spin, or z-spin. A bulb on a detector lights to indicate whether the spin-component is measured to be (+) or (−).

measured, the wave-function collapses accordingly, and so the particle is subsequently in whatever pure x-spin state corresponds to the measurement outcome. Furthermore, QM specifies that a particle in a pure x-spin state cannot also be in a pure y-spin state or in a pure z-spin state. This restriction is a form of Heisenberg's "uncertainty principle" and prohibits simultaneously measuring two components of the same particle's spin (since, as I just mentioned, a particle whose x-spin has just been measured is in a pure x-spin state, and likewise for the measurement of any other spin component). Finally, QM specifies that for any given pair emitted by this special source, if one particle's x-spin is measured and the outcome is (+), so that the wave-function's collapse leaves that particle's state as |x+ >, then its collapse leaves the other particle's state as |x− >, and vice versa. (Analogous remarks apply to y-spins and z-spins.) Prior to any spin measurement, then, the relevant "part" of the particle pair's wave-function is

$$(\sqrt{0.5})|xL+ > |xR− > − (\sqrt{0.5})|xL− >|xR+ >,$$

where "L" refers to the left particle and "R" to the right particle. (The minus sign is irrelevant to our purposes.) All of the probabilities that QM predicts for this case have been well confirmed experimentally.

Now let us see what Einstein meant by "spooky actions at a distance." Suppose we measure the left particle's x-spin and get the (+) outcome. If QM is complete, then the left particle was not |x+ > before the measurement was made. Did the left particle's interaction with the measuring device cause it to become |x+ >? Apparently so (though later we will think more about this). This causal relation would at least satisfy spatiotemporal locality. In contrast, consider what happens on the right wing of the apparatus. The right particle must become |x− >

exactly when the left particle becomes |x+ >, whether or not the right particle is being measured at that moment. Apparently, either the left particle's interacting with the left measuring device or the left particle's becoming |x+ > causes the right particle to become |x− >. But the right particle is in contact with neither the left measuring device nor the left particle. So spatial locality is violated. Action at a distance! (I'll say more later about why it is especially "spooky"!)

Of course, if the particles were already |xL+ > and |xR− > in advance of their being measured, then the left measurement merely revealed the pre-existing state of the left particle and, by inference, of the right particle. Locality would not be violated; no change in the right particle would accompany the left particle's measurement. But QM would then be incomplete, since QM does not describe the particles as being in pure spin states before they are measured.

Once an x-spin measurement has been made on one wing of the apparatus, both particles are in pure x-spin states. A subsequent y-spin measurement on one wing does not require that the *other* wing's particle afterwards be in a pure y-spin state; the "connection" between the two particles has been severed by the first measurement. According to QM, once an x-spin measurement has been made on one wing, a later y-spin measurement on either wing is 50 percent likely to yield (+), and simultaneous y-spin measurements on the two wings are equally likely to yield (++), (+−), (−+), or (−−). If we simultaneously measure x-spin on the left and y-spin on the right, then of course, both measurements have definite outcomes. But upon measurement, it is no longer the case that |xL+ > must be accompanied by |xR− >. So once again, no arrangement produces a particle that is simultaneously in pure states of two spin components. (But all of this talk of simultaneity is problematic; see box 9.1.)

2   The Bell Inequalities

We have, then, a dilemma: Either QM is incomplete (it leaves out "hidden variables") or QM is complete and so spatiotemporal locality fails. Which is it? In a famous 1935 paper (Wheeler and Zurek 1983: 138–41), Einstein along with two junior colleagues, Boris Podolsky (1896–1966) and Nathan Rosen (1909–95), used an example along roughly the lines we have been examining in order to argue that QM is incomplete. (After the first letters of the authors' last names, their

---

Box 9.1
QM and relativity

If QM is complete, then when the first measurement on the particle pair is made (say, on the left wing of the apparatus), the wave-function must simultaneously collapse on the right wing. But as we saw in chapter 8, simultaneity is frame-dependent. Relativity therefore makes it difficult to interpret QM as complete.

Let us see why. Suppose that a system's wave-function completely describes it and QM supplies genuine laws of nature saying that the wave-function's collapse is instantaneous. If QM is interpreted as saying that the collapse is instantaneous only in some particular reference frame, then QM privileges that frame as the one in which the laws of nature apply; two events that are simultaneous in that frame are really simultaneous, and a body's velocity in that frame is its absolute velocity. But this conflicts with relativity's general principle that reality can be described accurately from any inertial frame. So let's interpret QM as saying that the wave-function's collapse is instantaneous in any inertial frame. Now suppose that the spatiotemporal interval $I = (c\Delta t)^2 - (\Delta s)^2$ separating the left and right measurement events is negative. Then in different inertial frames, the order in which these measurements take place is different. QM therefore describes the sequence of events differently in different frames. For instance, in a frame where the left measurement (of x-spin, say) occurs first, the right particle is measured (for its y-spin, say) when it is already in a pure x-spin state. On the other hand, in a frame where the right measurement occurs first, the right particle is not in a pure x-spin state before it is measured, but the left particle is in a pure y-spin state before it is measured. (Of course, the two frames agree on the measurement outcomes and on the fact that the experiment leaves the left particle in a pure x-spin state and the right particle in a pure y-spin state.) How can the wave-function be a complete description of the system when the wave-function specifies a different sequence of events in different frames? Admittedly, relativity theory says that certain facts (such as the separation in time or space between two events) are frame-dependent. But relativity also tells us about objective reality by specifying the values of various Lorentz invariant quantities (such as the spatiotemporal interval). QM supplies nothing obviously analogous.

Now suppose that both detectors are set to measure x-spin. Suppose the left result is (+) and so the right result must be (−). If a measurement causes a change in the measured particle and (contrary to locality) in

the other particle, then in the frame where the left measurement precedes the right, the left measurement causes the left particle to become
|x+ > and the right particle to become |x− >. But in the frame where the
right measurement precedes the left, the right measurement does the
causing. Causal relations are themselves frame-dependent. (We saw in
chapter 8 that this would not happen as long as causal influences
traveled exclusively at c.)

Thus, relativity makes it difficult to interpret wave-function collapse
as a genuine physical transformation. (For more on how wave-function
collapse could be understood relativistically, see Maudlin 1994.) But we
are about to see obstacles to interpreting wave-function collapse as
merely a change in our knowledge of the system.

argument is commonly termed the "EPR argument" or the "EPR
paradox.") However, to most physicists in the middle of the twentieth
century, this dilemma – incompleteness or nonlocality – was not a
burning issue. QM was (and is) extraordinarily successful at predicting
observations. Whether a particle possesses a definite component spin
prior to its being measured, or instead acquires a definite component
spin only upon being measured, would apparently make no difference
to the outcome of any measurement: in either case, the measurement
would have a definite outcome. We obviously cannot *measure* whether
a particle possesses some property prior to its being measured, because
whenever we measure the particle, we are no longer dealing with the
particle prior to its being measured.

However, one of the lessons we have learned in this book (especially
in the final sections of chapters 5 and 8) is that even if a fact is not
detectable directly, and no matter how remote it seems from anything
that can be measured, we cannot know for sure that further theoretical
work will not establish a relation between that fact and something directly
observable, thereby allowing the fact to be ascertained. Perhaps, then,
an observable consequence can be coaxed out of the idea that a particle has a definite component spin prior to its being measured. In 1964,
someone did just that: John Bell (1928–90), a physicist from Belfast
working at the European Center for Nuclear Research (CERN) near
Geneva. (His PhD dissertation, by the way, was directed at the University of Birmingham in 1955 by Peierls, who apparently never succeeded
in making Bell satisfied with the Copenhagen interpretation.)

I now present a "quick and dirty" version (after d'Espagnat 1979) of Bell's original argument (Wheeler and Zurek 1983: 403–8). In the next section (which may be skipped without loss of continuity), I give Bell's argument in greater technical detail and with a slightly more careful account of the assumptions on which it is based. To tease out these assumptions has been the aim of many papers, since the upshot is that at least one of them must be mistaken.

Suppose that QM is incomplete: the two particles in the apparatus in figure 9.1 are characterized by hidden variables. These properties could be anything, of course, but to keep things simple, let us suppose that the particles have definite spin components in all directions prior to being measured. The key fact is that the particles' being characterized by these hidden variables, whatever they are, would be *local* causes of the measurement outcomes on the two wings, upholding spatiotemporal locality. Because of the special source being used, the left particle has spin components x+, y−, and z+ if and only if the right particle in the pair has spin components x−, y+, and z−. Let N(x+,y−,z+) be the number of particle pairs emitted during a certain period where the left particle possesses spin components x+, y−, and z+. Likewise, let N(x+,y−) be the number of pairs emitted during that period where the left particle is x+ and y−; the particle's z-spin does not matter. Now obviously, if the particles have definite spin components prior to their being measured, then a particle that is x+ and y− is either x+, y−, and z+ or x+, y−, and z−. So

$$N(x+,y-) = N(x+,y-,z+) + N(x+,y-,z-). \qquad (9.1)$$

Likewise,

$$N(x+,z-) = N(x+,y+,z-) + N(x+,y-,z-). \qquad (9.2)$$

It follows from (9.2) that N(x+,z−) cannot be less than N(x+,y−,z−) since any particle that is x+, y−, z− is automatically x+, z−. So

$$N(x+,z-) \geq N(x+,y-,z-). \qquad (9.3)$$

Likewise,

$$N(y-,z+) = N(x+,y-,z+) + N(x-,y-,z+) \geq N(x+,y-,z+). \qquad (9.4)$$

Substituting (9.3) and (9.4) into (9.1) yields

$$N(x+,y-) \leq N(y-,z+) + N(x+,z-). \qquad (9.5)$$

Now of course, we cannot measure $N(x+,y-)$ directly, since according to QM, we cannot measure simultaneously the same particle's x-spin and y-spin. (Whether this prohibition holds because the laws of nature prohibit certain kinds of apparatus from being made, just as they prohibit perpetual-motion machines and particle accelerators that accelerate particles beyond the speed of light, or whether this limitation holds because a given particle cannot *possess* a definite x-spin along with a definite y-spin, depends on whether or not QM is complete.) But if particles have definite spins prior to being measured (so that QM is incomplete) and the measurements simply reveal these pre-existing spin values, then (because the particles in a given pair are correlated) $N(x+,y-)$ is the number of pairs measured to be xL+ and yR+. So (9.5) entails

$$N(xL+,yR+) \leq N(yL-,zR-) + N(xL+,zR+). \qquad (9.6)$$

Again, we cannot measure $N(xL+,yR+)$ at the same time as we measure $N(yL-,zR-)$, since we cannot measure two spin components of the same particle simultaneously. But let's assume that for every spin combination (such as xL+, yR+), there is a certain probability that the source will emit a pair of particles with that spin combination, a probability unaffected by whether or not we choose to measure those spin components on the two wings of the apparatus. Suppose we begin by setting our detectors simply to detect every particle pair that is emitted during a certain time interval Δt (without measuring their spins). Suppose that over many repetitions of this experiment, p is the average number of particles emitted during Δt. Suppose we then set our detectors to measure xL and yR, and over many repetitions of this experiment, q is the average $N(xL+,yR+)$ during Δt. Then as the number of repetitions becomes greater or Δt becomes larger, q/p becomes arbitrarily likely to be $pr(xL+,yR+)$: the probability that an emitted pair will be $(xL+,yR+)$. We can then reset our detectors to measure yL and zR, and thus ascertain $pr(yL-,zR-)$. If we assume that the behavior of the source is unaffected by what properties we have set the detectors to measure, we can regard our measured $pr(xL+,yR+)$ and $pr(yL-,zR-)$ as probabilities applying to the source at the same moment, even though they were measured on different runs of the experiment. It then follows from (9.6) that

$$pr(xL+,yR+) \leq pr(yL-,zR-) + pr(xL+,zR+). \qquad (9.7)$$

This is one of the "Bell inequalities."

Remarkably, QM's predictions regarding the probabilities in (9.7) can conflict with (9.7). This dooms any attempt to develop a theory that predicts the same probabilities as QM does, but describes the world in terms of hidden variables obeying the commonsensical assumptions used to derive the Bell inequalities. These assumptions include spatiotemporal locality; the hidden variables were intended to be *local* causes of the measurement outcomes. (In the next section, I shall work on making Bell's other assumptions more explicit.) In other words, Bell's argument shows that to include local hidden variables obeying Bell's other assumptions, a theory must depart from QM's predictions regarding our observations, since the theory's predictions must obey Bell's inequalities, which QM does not.

Notice that Bell's argument is *not* that scientists tried for many years to come up with a local hidden-variable theory obeying Bell's other assumptions and reproducing QM's probabilities, but each of their candidates failed, so since scientists are smart, it is probably not possible to formulate such a theory. This would be a relatively weak argument. It would be like arguing that since no one has ever managed to come up with a triangle on a Euclidean plane whose interior angles fail to total 180 degrees, even though many smart people have tried, such a triangle must be impossible. In geometry, you can give a better argument than this: you can *prove* in advance of any trial-and-error that such a triangle is impossible, simply by virtue of what such a triangle would have to be like. In the same way, Bell's argument *proves* that a certain inequality must be obeyed by every local hidden-variable theory obeying certain other assumptions: every such theory that has already been formulated, that eventually will be formulated, or that no one will ever formulate. It is remarkable that all of these theories, no matter how different in their details, have something non-trivial in common. Yet they do: the Bell inequalities. And what's more, these inequalities are so non-trivial that QM violates them!

From 1972 to 1985, a dozen or so experiments were carried out involving devices of the sort we have been discussing. Their over-whelming verdict was that QM's predictions are accurate (Clauser and Shimony 1978; Redhead 1987: 107); Bell's inequalities are false. (This did not surprise most physicists, considering the tremendous experimental support that QM had already received.) It follows from Bell's argument, then, that even if QM is incomplete in that it fails to describe certain properties that particles possess, these hidden variables must violate one of Bell's assumptions. For example, the source's probability of emitting a particle pair with certain definite spin components (or whatever the hidden variables may be) was assumed to be unaffected by whether the devices are set to measure x-spin, y-spin, or z-spin. Suppose this assumption is mistaken. Then QM's predictions could be accurate (violating Bell's inequalities) even while the measurement outcomes are caused locally. But how could a measuring device's setting influence the distant source's behavior without violating spatial locality? Alternatively, suppose that the particles possess definite spin components prior to their being measured, but the outcome of a left spin measurement affects the spin of the right particle. Then once again, QM's predictions could be accurate (violating Bell's inequalities) even while the measurement outcomes are caused locally. But spatial locality is again violated in the left measurement outcome's affecting the spin components of the right particle (if no causally relevant events take place in the gap between the left measurement and the right particle).

In short, Einstein's argument (in the previous section) left us with a dilemma: either QM is incomplete, leaving room for hidden variables that obey locality, or QM is complete and spatiotemporal locality is violated, as a measurement on the left particle nonlocally affects the particle on the right. Bell's argument shows that as long as the probabilities predicted by QM match nature, there is no room for hidden variables obeying all of Bell's assumptions, notable among which is spatiotemporal locality. Unless one of Bell's other assumptions can take the blame for experimental violations of Bell's inequalities, *neither* side of Einstein's dilemma will permit spatiotemporal locality to hold.

What assumption could take the blame if not locality? Well, for a start, I have left out various refinements of Bell's inequalities that had to be made in order for them to be tested experimentally. For instance, an *actual* device for detecting pairs of particles will not function with perfect efficiency. Sometimes, the spin component of one particle in a pair will be measured but the other particle's spin component will not be detected. (For example, a photon detector typically detects only 10 percent of the photons passing through it. So on average, only 1 out of every 100 pairs of photons will have both of its members measured.) The experimenter must then decide what to do with the incomplete pieces of data. It might seem reasonable simply to discard them in the belief that the device was malfunctioning when they were recorded and that had it been working, these data would have turned out to comprise a random sample of all of the data collected. It is possible, however, that particles with certain hidden-variable values are more likely to be missed by the detectors. Then the detected particles would likely form a biased sample of what the source emits. Thus, our discarding the incomplete data may result in the *observed* N(xL+,yR+) systematically deviating from the *actual* pr(xL+,yR+). In fact, it has been shown (Pearle 1970) that if all incomplete data were disregarded by an experimenter, certain hidden-variable theories obeying spatiotemporal locality and Bell's other assumptions would agree experimentally with QM's predictions, violating Bell's inequalities. This "detection loophole" can be circumvented by making a few extra technical assumptions that I'll not go into – very plausible, but not utterly unquestionable assumptions (Clauser and Horne 1974; Fine 1991). Experimental violations of Bell's inequalities could be blamed on the failure of these assumptions rather than of locality.[2]

One of the most noteworthy experimental tests of Bell's inequalities was performed by Alain Aspect and his colleagues at the University of Paris's Institute for Theoretical and Applied Optics (Aspect et al. 1981; Shimony 1988). They found QM's predictions to be accurate, violating Bell's inequalities, even when the left and right detectors are set (for example, to measure x-spin and y-spin, respectively) *after* the two particles have been emitted from the source – and when they are so close to their respective detectors that a signal traveling at the speed of light and beginning at the moment that one detector is set could not reach the other wing before the particle there is measured. This makes it implausible that the left detector's being set affects the right measurement outcome. Moreover, unless an effect can somehow precede its cause, the Aspect experiment also rules out the possibility that Bell's inequalities are violated because the detectors' being set causes a change in the source's probabilities of emitting particles with various properties.[3] It is possible, of course, that some obscure past event locally set in motion causal chains that ultimately caused both the detectors' settings and the source's behavior, coordinating them in such a way as to violate Bell's inequalities. A loophole in the Aspect experiment from which such a "conspiracy" theorist could take heart arises from the fact that the detectors were not truly set randomly. But a group at the University of Innsbruck has recently attempted to close this loophole (Weihs et al. 1998).

Experiments like Aspect's make it tremendously tempting to conclude that some sort of nonlocal causal influences are at work. In sections 4 and 5, we shall grapple with this possibility.

3   For Whom the Bell Tolls

What are the barest assumptions sufficient to derive Bell's inequalities? We do not need to assume the strict correlation given in the previous section: that the left particle's state is |x+ > if and only if the right particle's state is |x− >, and vice versa (and likewise for their y-spins and z-spins). Furthermore, we do not need to assume that the hidden variables are spin components or any other properties familiar from QM, nor need we assume that a particle's value λ of some hidden variable *determines* the outcome of, say, an x-spin measurement on that particle. It is enough that the hidden-variable's value determines the *probability* of the measurement outcome, given the measurement device's setting. The hidden variable would then still underwrite spatiotemporal locality. Let us now derive Bell's inequalities using these weaker assumptions.

Suppose you toss a coin repeatedly and the outcome is heads, tails, tails, heads, heads, heads, ... If we take heads to be +1 and tails to be −1, then our sequence is +1, −1, −1, +1, +1, +1 . . . . At each point, we can take the average of the outcomes so far: +1, 0, −$\frac{1}{3}$, 0, +$\frac{1}{5}$, +$\frac{1}{3}$ . . . . The sequence of averages will almost certainly home in on some particular value (so long as there is no change from toss to toss in the mass distribution within the coin). For instance, if the coin is "fair," then the sequence of averages will almost certainly in the long run approach zero arbitrarily closely. This is called the coin toss's "expectation value." In other words, the "expectation value" of an observable quantity is what the average of repeated measurements of that quantity approaches in the long run. The expectation value is a kind of *weighted average* of the various possible measurement outcomes, an outcome's weight being its probability. For instance, if the coin is fair, then pr(heads) = pr(tails) = 0.5, and so the expectation value = (+1) pr(heads) + (−1) pr(tails) = 0.

Now suppose that for any particle pair emitted by the source, there is a value λ of some hidden variable. One of Bell's assumptions is that λ and the settings of the two measuring devices determine the probabilities of getting various combinations of measurement outcomes on the two wings. For instance, let "pr(+−|λLR)" stand for the probability of getting outcome (+) on the left device and (–) on the right, given that the particle pair is characterized by λ, the left device's setting is L, and the right device's setting is R. (Each device can measure one of two different spin components: L or L′ on the left, and R or R′ on the right.) Let "pr($+_{\text{left}}$|λLR)" stand for the probability of getting outcome (+) on the left wing, given that the pair is characterized by λ and the device's settings are L and R. Let us think of the (+) outcome as +1 and the (–) outcome as –1. Then we can define "E_{left}(λLR)" as the expectation value of the left measurement outcome when λ characterizes the particle pair and the settings of the left and right devices are L and R, respectively. Just as the expectation value of the coin flip was the weighted average (+1)pr(heads) + (–1)pr(tails), so

$$E_{\text{left}}(\lambda LR) = (+1)\text{pr}(+_{\text{left}}|\lambda LR) + (-1)\text{pr}(-_{\text{left}}|\lambda LR)$$

$$= \text{pr}(+_{\text{left}}|\lambda LR) - \text{pr}(-_{\text{left}}|\lambda LR). \tag{9.8}$$

Instead of considering just a left-wing (or right-wing) measurement, we could consider the measurements on the two wings together by multiplying the left outcome (+1 or –1) by the right outcome (+1 or –1). The result is +1 if the two outcomes are the same and –1 if they are different. Here is the expectation value of that combined measurement:

$$E_{\text{lr}}(\lambda LR) = (+1)\text{pr}(++|\lambda LR) + (+1)\text{pr}(--|\lambda LR) + (-1)\text{pr}(+-|\lambda LR)$$
$$+ (-1)\text{pr}(-+|\lambda LR)$$

$$= \text{pr}(++|\lambda LR) + \text{pr}(--|\lambda LR) - \text{pr}(+-|\lambda LR)$$
$$- \text{pr}(-+|\lambda LR). \tag{9.9}$$

Suppose that the source emits particle pairs with various values of the hidden variable. Let pr(λ|LR) be the probability of an emitted pair's hidden variable having the value λ, given that the two devices are set on L and R, respectively. One of Bell's assumptions is that the probability of an emitted pair's hidden variable having the value λ is independent of the settings of the two devices. So we may simply write "pr(λ)" rather than "pr(λ|LR)." Bell thus assumes that the act of setting the detectors does not somehow cause a change in the distant source's tendency to emit particles with various properties. Bell's assumption also precludes the "conspiracy" theory: that some obscure past event locally set in motion causal chains that ultimately caused both the detectors' settings and the source's behavior, coordinating them.

Now consider $E_{left}(LR)$, the expectation value of the left measurement outcome with the devices set on L and R. This is a weighted average of the various possible left outcomes (+1 and −1), each weighted by its probability of occurring with the devices set on L and R respectively:

$$E_{left}(LR) = (+1)pr(+_{left}|LR) + (-1)pr(-_{left}|LR). \qquad (9.10)$$

The probability of getting +1 on the left is equal to the probability of getting +1 on the left when the particle pair is characterized by λ times the probability of its being characterized by λ (that is, $pr(+_{left}|\lambda LR)$ $pr(\lambda)$, using $pr(\lambda|LR) = pr(\lambda)$ as we just assumed) plus the same thing for some other value of the hidden variable, and so forth for all of the hidden variable's possible values. Since the hidden variable has a continuum of possible values, we must express this sum as an integral over all possible values λ of the hidden variable:

$$pr(+_{left}|LR) = \int pr(+_{left}|\lambda LR)pr(\lambda)d\lambda. \qquad (9.11)$$

Substituting (9.11) and the analogous result for $pr(-_{left}|LR)$ into (9.10), we find

$$\begin{aligned}
E_{left}(LR) &= \int [(+1)pr(\lambda)pr(+_{left}|\lambda LR) + (-1)pr(\lambda)pr(-_{left}|\lambda LR)]d\lambda \\
&= \int pr(\lambda)[pr(+_{left}|\lambda LR) - pr(-_{left}|\lambda LR)]d\lambda \\
&= \int pr(\lambda)E_{left}(\lambda LR)d\lambda, \qquad (9.12)
\end{aligned}$$

where the final step uses (9.8). By the same reasoning, but using (9.9) rather than (9.8), we find

$$E_{lr}(LR) = \int pr(\lambda)E_{lr}(\lambda LR)d\lambda. \qquad (9.13)$$

These integrals assume that particle pairs with different values of the hidden variable are measured with equal accuracy. (Otherwise the differences in detector sensitivity must be weighed into the expectation value.) Here, then, is another of Bell's assumptions (which I also mentioned in the previous section).

Let us return to the coin being tossed. In a sense, the coin has a hidden variable: its bias – that is, whether it is fair, or biased a bit towards heads, or very biased towards tails, or whatever. Suppose we flip the coin twice. Getting a head on the first flip makes it a bit more probable that the coin is biased towards heads than towards tails, and therefore makes it a little more likely that the second flip will also land heads. That is, the outcome of the second toss is a little more likely to be heads when the first toss landed heads than when the first toss landed tails:

$$pr(head2|head1) > pr(head2|tail1). \qquad (9.14)$$

In other words, the probability of a head outcome on the second toss is greater given that the first toss lands heads:

$$pr(head2|head1) > pr(head2). \qquad (9.15)$$

(You can read "pr(head2|head1)" as "the probability that the second toss lands heads given that the first toss lands heads," whereas "pr(head2)" is the probability that the second toss lands heads where nothing is specified regarding the first toss's outcome; it could have been heads or tails.) Equation (9.14) holds exactly when (9.15) holds, and they in turn hold exactly when

$$pr(head1 \text{ and } head2) > pr(head1) \, pr(head2). \qquad (9.16)$$

(See box 9.2.) Another way of putting (9.14), (9.15), and (9.16) is that the outcomes of the two tosses are not *statistically independent*. But as everyone knows, their statistical independence is restored once the bias of the coin (the "hidden variable") is specified. For example, given that the coin is fair, the chance of the second toss landing heads is 50 percent regardless of what the first toss's outcome was. In other words:

$$pr(head2|head1 \text{ and } bias) = pr(head2|bias)$$

$$pr(head1 \text{ and } head2|bias) = pr(head1|bias) \, pr(head2|bias).$$
$$(9.17)$$

For instance, given that the coin is biased 75 percent towards heads, the probability of getting heads on the second toss is 75 percent whether or not the first toss landed heads, and the probability of getting two heads in a row is $(0.75)(0.75) = \frac{9}{16}$.

This is an instance of a general principle. Suppose we have two kinds of events, A and B, where A neither causes B nor is caused by B, and yet $pr(B|A) > pr(B)$, as in (9.15), or equivalently, $pr(A \text{ and } B) >$

Box 9.2
Probabilities, conditional and unconditional

To see why (9.14), (9.15), and (9.16) are all "logically equivalent" (that is, they must either all be true or all be false), we must turn to some basic facts about probabilities. Pr(head2|head1) is a "conditional probability." The general rule relating conditional to unconditional probabilities is

$$pr(A|B) = pr(A \text{ and } B)/pr(B)$$

so long as pr(B) is non-zero. (Let's just say that pr(A|B) is undefined when pr(B) = 0.) In other words, A's probability given B is the fraction of B's probability that also contributes to A's probability. For example, a weather forecaster might say that rain is 75 percent likely today considering the current barometric reading:

$$0.75 = pr(\text{rain today}|\text{barometer reading})$$
$$= pr(\text{rain today and barometer reading})/$$
$$pr(\text{barometer reading}).$$

Perhaps pr(barometer reading) = 12 percent and pr(rain today and barometer reading) = 9 percent, so that given that we are in the former 12 percent, it is 75 percent likely that we are also in the latter 9 percent. In our original example,

$$pr(\text{head2}|\text{head1}) = pr(\text{head1 and head2})/pr(\text{head1}).$$

This result allows us to go immediately between (9.15) and (9.16). (You might try proving for yourself that pr(A|B) = pr(A) exactly when pr(B|A) = pr(B).)

Here is another basic fact about probabilities:

$$pr(A \text{ and } B) + pr(\text{notA and } B) = pr(B).$$

That is because there are only two ways for B to hold, and these two ways exclude each other: with A holding and with A not holding. For example, since the first toss lands either heads or tails but not both,

$$pr(\text{head1 and head2}) + pr(\text{tail1 and head2}) = pr(\text{head2}).$$

Now as we just saw,

$$pr(head1 \text{ and } head2)/pr(head1) = pr(head2|head1),$$

and

$$pr(tail1 \text{ and } head2)/pr(tail1) = pr(head2|tail1).$$

Substituting these last two results into the previous equation yields

$$pr(head2|head1) \, pr(head1) + pr(head2|tail1) \, pr(tail1) = pr(head2).$$

So pr(head2) is a weighted average of head2's probabilities given the two possible outcomes of the first toss.

Hence, (9.14) and (9.15) are logically equivalent. To see why, start with (9.15) and substitute for pr(head2):

$$pr(head2|head1) > pr(head2|head1)pr(head1) + \\ pr(head2|tail1)pr(tail1).$$

So

$$pr(head2|head1) - pr(head2|head1)pr(head1) > \\ pr(head2|tail1)pr(tail1).$$

Or

$$pr(head2|head1)[1 - pr(head1)] > pr(head2|tail1)pr(tail1).$$

But

$$pr(head1) + pr(tail1) = 1$$

since the coin is 100 percent likely to land either heads or tails and these two outcomes are mutually exclusive. So

$$pr(head2|head1)pr(tail1) > pr(head2|tail1)pr(tail1).$$

Therefore, as long as pr(tail1) is non-zero (as it must be for pr(head2|tail1) to be well defined)

$$pr(head2|head1) > pr(head2|tail1).$$

That's (9.14).

$pr(A)pr(B)$, as in (9.16). The two events' statistical independence is restored if we take into account their *common causes*. (The bias of the coin is the common cause of the outcomes of the two flips.) If C contains all of A's causes that are also B's causes, then $pr(B|A$ and $C) = pr(B|C)$ and $pr(A$ and $B|C) = pr(A|C)$ $pr(B|C)$. Their common cause C "screens A off from B" – makes A and B statistically independent of each other.

Now return to the two-wing apparatus. One of Bell's assumptions is that λ is the only common cause of the outcomes on the two wings. Hence, by the "screening off" principle,

$$pr(+-|\lambda LR) = pr(+_{\text{left}}|\lambda LR)\, pr(-_{\text{right}}|\lambda LR). \tag{9.18}$$

In other words, once λ is given, the results on the two wings are statistically independent of each other: $pr(+_{\text{left}}|-_{\text{right}}\lambda LR) = pr(+_{\text{left}}|\lambda LR)$.

It is also assumed that λ and L contain all of the causes of the outcome on the left wing (and analogously for λ and R on the right wing). Hence, each outcome is caused *locally*, since λ and L are local to the left measurement event. This suggests Bell's final assumption:

$$pr(+_{\text{left}}|\lambda LR) = pr(+_{\text{left}}|\lambda L)$$
$$pr(-_{\text{right}}|\lambda LR) = pr(-_{\text{right}}|\lambda R). \tag{9.19}$$

Equations along these lines are sometimes called "parameter independence."[4] Substituting (9.19) into (9.18), we find

$$pr(+-|\lambda LR) = pr(+_{\text{left}}|\lambda L)pr(-_{\text{right}}|\lambda R). \tag{9.20}$$

Let us now use equations of this form (often called "factorizability") to compute the expectation value of the combined left–right measurement. We had already found

$$E_{\text{lr}}(\lambda LR) = pr(++|\lambda LR) + pr(--|\lambda LR) -$$
$$pr(+-|\lambda LR) - pr(-+|\lambda LR). \tag{9.9}$$

By factorizability,

$$E_{\text{lr}}(\lambda LR) = pr(+_{\text{left}}|\lambda L)pr(+_{\text{right}}|\lambda R) + pr(-_{\text{left}}|\lambda L)pr(-_{\text{right}}|\lambda R) -$$
$$pr(+_{\text{left}}|\lambda L)pr(-_{\text{right}}|\lambda R) - pr(-_{\text{left}}|\lambda L)pr(+_{\text{right}}|\lambda R). \tag{9.21}$$

Combining like terms:

$$E_{lr}(\lambda LR) = pr(+_{left}|\lambda L)[pr(+_{right}|\lambda R) - pr(-_{right}|\lambda R)] -$$
$$pr(-_{left}|\lambda L)[pr(+_{right}|\lambda R) - pr(-_{right}|\lambda R)]. \qquad (9.22)$$

Combining like terms again:

$$E_{lr}(\lambda LR) = [pr(+_{left}|\lambda L) - pr(-_{left}|\lambda L)][pr(+_{right}|\lambda R) - pr(-_{right}|\lambda R)]. \qquad (9.23)$$

In (9.8), we saw that $E_{left}(\lambda LR) = pr(+_{left}|\lambda LR) - pr(-_{left}|\lambda LR)$. By "parameter independence" (9.19), it follows that $E_{left}(\lambda LR) = pr(+_{left}|\lambda L) - pr(-_{left}|\lambda L)$, which we might just as well call $E_{left}(\lambda L)$ since it doesn't depend on R. Substituting into (9.23), we arrive at

$$E_{lr}(\lambda LR) = [E_{left}(\lambda L)][E_{right}(\lambda R)]. \qquad (9.24)$$

Let us use this result to derive Bell inequalities (following Bell 1971). Substituting (9.24) into our earlier result

$$E_{lr}(LR) = \int pr(\lambda)E_{lr}(\lambda LR)d\lambda, \qquad (9.13)$$

we find

$$E_{lr}(LR) = \int pr(\lambda)E_{left}(\lambda L)E_{right}(\lambda R)d\lambda, \qquad (9.25)$$

and likewise for a different setting of the right detector,

$$E_{lr}(LR') = \int pr(\lambda)E_{left}(\lambda L)E_{right}(\lambda R')d\lambda. \qquad (9.26)$$

So

$$E_{lr}(LR) - E_{lr}(LR') = \int pr(\lambda)E_{left}(\lambda L)E_{right}(\lambda R)d\lambda$$
$$- \int pr(\lambda)E_{left}(\lambda L)E_{right}(\lambda R')d\lambda. \qquad (9.27)$$

Let us subtract and add the same quantity: $\int pr(\lambda)E_{left}(\lambda L)E_{right}(\lambda R)$ $E_{left}(\lambda L')E_{right}(\lambda R')d\lambda$, for another setting L' of the left detector. This gives us

$$E_{lr}(LR) - E_{lr}(LR') = \int pr(\lambda)E_{left}(\lambda L)E_{right}(\lambda R)d\lambda$$
$$- \int pr(\lambda)E_{left}(\lambda L)E_{right}(\lambda R)E_{left}(\lambda L')E_{right}(\lambda R')d\lambda$$
$$- \int pr(\lambda)E_{left}(\lambda L)E_{right}(\lambda R')d\lambda$$
$$+ \int pr(\lambda)E_{left}(\lambda L)E_{right}(\lambda R)E_{left}(\lambda L')E_{right}(\lambda R')d\lambda. \qquad (9.28)$$

Desperately combining terms before things get really out of hand:

$$E_{lr}(LR) - E_{lr}(LR')$$
$$= \int pr(\lambda)E_{left}(\lambda L)E_{right}(\lambda R)[1 - E_{left}(\lambda L')E_{right}(\lambda R')]d\lambda$$
$$- \int pr(\lambda)E_{left}(\lambda L)E_{right}(\lambda R')[1 - E_{right}(\lambda R)E_{left}(\lambda L')]d\lambda. \quad (9.29)$$

Since each of these expectation values is a weighted average of 1 and −1, it must be between 1 and −1. So each of the terms inside square brackets ranges from [1 − 1] to [1 − (−1)] – that is, from 0 to 2. Thus, both of the terms inside square brackets are non-negative. So to maximize the right side of (9.29), maximize the first term on the right side by making $E_{left}(\lambda L) \, E_{right}(\lambda R)$ equal 1 and minimize the second term by making $E_{left}(\lambda L) \, E_{right}(\lambda R')$ equal −1. Hence

$$E_{lr}(LR) - E_{lr}(LR') \leq \int pr(\lambda)[1 - E_{left}(\lambda L')E_{right}(\lambda R')]d\lambda$$
$$+ \int pr(\lambda)[1 - E_{right}(\lambda R)E_{left}(\lambda L')]d\lambda. \quad (9.30)$$

Since the integral is being taken over all possible values of the hidden variable and the probabilities of all these values must sum to 1, $\int pr(\lambda)d\lambda = 1$. So

$$E_{lr}(LR) - E_{lr}(LR') \leq 2 - \int pr(\lambda)E_{left}(\lambda L')E_{right}(\lambda R')d\lambda$$
$$- \int pr(\lambda)E_{right}(\lambda R)E_{left}(\lambda L')d\lambda. \quad (9.31)$$

Analogous to (9.25), we have

$$E_{lr}(L'R') = \int pr(\lambda)E_{left}(\lambda L')E_{right}(\lambda R')d\lambda, \text{ and}$$
$$E_{lr}(L'R) = \int pr(\lambda)E_{left}(\lambda L')E_{right}(\lambda R)d\lambda. \quad (9.32)$$

Substituting (9.32) into (9.31), we find at last

$$E_{lr}(LR) - E_{lr}(LR') \leq 2 - E_{lr}(L'R') - E_{lr}(L'R). \quad (9.33)$$

This is one of Bell's inequalities. For certain detector settings, it fails to hold according to QM.

Experiments have vindicated QM's predictions against Bell's inequality. So one of the assumptions used to derive (9.33) must be mistaken. As we saw, it is very tempting to conclude that some kind of nonlocal causal influence is operating.

4   Wrestling with Nonlocality

Even if locality is violated according to QM, we cannot exploit these "spooky actions at a distance" to send messages instantaneously.[5] Suppose you decided to try: you (located on the left wing of the apparatus) and your partner (on the right) arrange in advance the following code: when the next pair of particles is emitted from the source and you wish to send the message "yes," then you will measure the x-spin of the left particle. This will instantly collapse the wave-function of the right particle into a pure x-spin state. If you wish to send the message "no," then you will make no measurement of the left particle, and so the right particle will remain in a superposition. So all that your partner needs to do in order to receive your message is to check the particle on the right!

But it is not so easy: your partner cannot ascertain the state of the right particle except by measuring it. But if your partner uses an x-spin detector, then regardless of what you did on the left side, your partner's measurement will have a definite outcome. There is no way for your partner to ascertain whether the right particle was in a pure x-spin state prior to being measured by her, or whether her measurement brought it into a pure x-spin state.

Suppose you arrange a different code: if you wish to send the message "yes," then you will measure the x-spin of the left particle, whereas if you wish to send the message "no," then you will measure its y-spin. But this works no better. If your partner checks the x-spin of the right particle, then regardless of what you did, your partner's measurement will have a definite outcome. On the other hand, if your partner checks its y-spin, then regardless of what you did, your partner's measurement will have a definite outcome.

Of course, your code cannot be that if you wish to send the message "yes," then you will measure the x-spin of the left particle to be (+), whereas if you wish to send the message "no," then you will measure the x-spin of the left particle to be (–). This code cannot be used because you have no control over what outcome your measurement will have. There is simply a 50 percent chance that it will be (+) and a 50 percent chance that it will be (–). Thus, an instantaneous "Bell telephone" (John, not Alexander Graham) seems impossible. The measurement event on the left wing cannot be used to send information to the right wing. (Indeed, the probability of a right measurement's returning (+) is the same regardless of whether the left device measures x-spin or y-spin.)

Nevertheless, the left measurement event apparently helps to cause something to happen on the right (namely, the right particle's acquiring some definite spin component) with no causes in between – in violation of spatial locality. In light of earlier chapters, you might wonder whether we could restore locality by introducing some new kind of field. Perhaps we could, but the new field would have to be very different from the electromagnetic field (or Newtonian gravitational field). For one thing, the nonlocal influence is not weakened by distance; it makes no difference how far apart the two measuring devices are. An experiment performed by a group from the University of Geneva found no weakening of the correlation between spin measurements on the two wings up to a separation of seven miles (Tittel et al. 1999). This is obviously quite different from, say, electrostatic influences, which diminish with the square of the distance. Furthermore, we can shield one charged particle from the electrostatic influence of another by surrounding it with a conductor. No shielding can affect the quantum influence. A final important difference (apart from the obvious fact that the quantum influence travels instantaneously, not merely at the speed of light) is that the quantum influence is extremely selective. Unlike the electric field, which influences any charged body, measurement of the left particle affects only the other particle that was created along with it. Even if we have a million such experiments running simultaneously, only particles that were created together (no matter how long ago that was) can influence each other in this way. This, then, is not mere action at a distance, but "spooky action at a distance"!

To gain some perspective, let us look at another example of apparent nonlocality in QM. Suppose a source fires electrons at a barrier with two slits in it. Behind the barrier is a screen that indicates where an electron hits it (figure 9.2). Some locations on the screen are more likely to be hit than others. If we turn the source way down so that, most of the time, there is no more than a single electron in the apparatus, then we can watch a pattern on the screen build up from individual hits (figure 9.3). This pattern reflects the probabilities at various locations that an electron will hit the screen there. QM predicts these probabilities but does not determine with 100 percent certainty where any particular electron will go.

It might be supposed, however, that QM is incomplete – that an electron has hidden variables determining its trajectory. But whether QM is complete or there are hidden variables, an electron's state

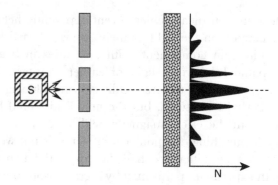

Figure 9.2   Electrons emitted by source S pass through the two slits of the barrier and then travel to the detector. The distribution of hits at the detector is shown behind it by the black curve; a higher number N indicates more hits.

seems to be subject to nonlocal influences. That is because the pattern produced on the screen (reflecting the probability of an electron's hitting it at various locations) changes in a peculiar way depending upon whether both of the barrier's slits are open or only one is open.

Let us analyze the situation under the assumption that an electron is characterized by a hidden variable giving it a definite trajectory through the apparatus. An electron can hit the screen only by passing through the top slit or by passing through the bottom slit. So the probability of a hit at a given location on the screen, if both slits are open, equals the probability that the source will emit an electron that passes through the top slit and then travels on to hit the screen at that location plus the probability that the source will emit an electron that passes through the bottom slit and then goes on to hit the screen there. (In the same way, in a race between Jones, Smith, and Brown, the probability that Jones wins is the probability that Jones wins and Smith comes in second plus the probability that Jones wins and Brown comes in second.) If "pr(hit|2)" stands for the probability of a hit at a given location on the screen if both slits are open, then

$$\text{pr(hit|2)} = \text{pr(hit from bottom|2)} + \text{pr(hit from top|2)}. \quad (9.34)$$

Now the probability that the source will emit an electron that passes through the *bottom* slit and then hits the screen at the given location should presumably be unaffected by whether the *top* slit is open or closed. (This is like Bell's assumption that $\text{pr}(\lambda|LR)$ can be represented simply as $\text{pr}(\lambda)$.) After all, as long as spatial locality holds, how would

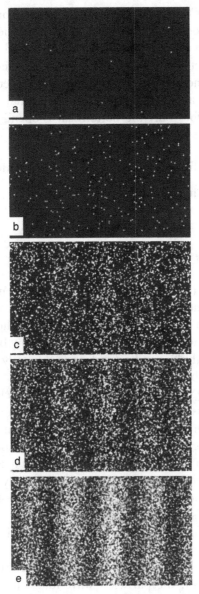

Figure 9.3 Gradual build-up of the pattern of hits on the detector. Each electron hits in only a single spot. Electrons arrive at approximately 1,000 per second. There is almost never more than a single electron in the apparatus at any moment. The approximate number of recorded electrons in frames (a) through (e) are respectively 10, 100, 3,000, 20,000, and 70,000. Only a portion of the entire detector is shown. (Courtesy of Akira Tonomura, Hitachi Advanced Research Laboratory, Hatoyama, Japan.)

the source "know" whether the top slit is open or closed, so as to change its likelihood of emitting electrons toward the bottom slit? (We could even set the slits *after* the source has emitted its particle, along the lines of Aspect's experiment.) Likewise, an electron presumably cannot "know" whether the top slit is open or closed, so as to change course toward or away from it, as long as spatial locality holds. So letting "b" stand for "only the bottom slit is open" and "t" stand for "only the top slit is open," we have

$$\text{pr(hit from bottom}|2) = \text{pr(hit from bottom}|b) \qquad (9.35)$$

$$\text{pr(hit from top}|2) = \text{pr(hit from top}|t). \qquad (9.36)$$

Substituting (9.35) and (9.36) into (9.34), we find

$$\text{pr(hit}|2) = \text{pr(hit from bottom}|b) + \text{pr(hit from top}|t). \qquad (9.37)$$

But QM predicts (9.37) to be false, and experiments confirm QM's prediction. At some locations on the screen, a hit is *more* likely with only one slit open than with *both* slits open! At the location on the screen exactly between the two slits, pr(hit from bottom|b) = pr(hit from top|t), but pr(hit|2) can be *four times* (not merely twice) as great as either pr(hit from bottom|b) or pr(hit from top|t).

According to QM, the wave-function of a particle that has passed the barrier, but has not yet reached the screen, is a superposition of a pure state of having gone through the top slit and a pure state of having gone through the bottom slit. If we seek the particle just after it has passed through the barrier, we are 50 percent likely to find it in the neighborhood of the top slit and 50 percent likely to find it in the neighborhood of the bottom slit. The particle's wave-function collapses when the particle's position is detected; in the apparatus shown in figure 9.2, detection occurs at the screen. If instead we measure the particle's position earlier – say, as it passes through the slits – then we can ascertain which one of the two slits it passed through. As a result of this measurement, the particle's wave-function collapses before it reaches the screen. This collapse changes the probability of a particle's arriving at a given location on the screen. In other words, if we install particle detectors at the slits (and these detectors allow a detected particle to pass through them, heading toward the screen), then the pattern built up on the screen is not the pattern we got in our earlier experiment. Instead, it is a pattern that agrees with (9.37).

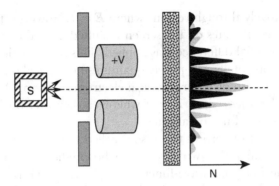

Figure 9.4 Apparatus for displaying the Aharanov–Bohm effect. Metallic cylinders are placed behind the two slits; the upper cylinder can be raised to a higher electric potential relative to the lower. The pattern of hits (black curve) is shifted as compared to the pattern built up when no potential difference is imposed (gray curve, as in figure 9.2).

In connection with the double-slit apparatus, QM adds something to an issue that we last discussed in chapter 2: whether the electric potential is real or merely a theoretical device. In classical electromagnetic theory, the potential's *absolute value* at a given location makes no difference to any other facts; it is a "dangler." Facts about potential *differences* seem to be made true by facts about the electric field. Therefore, the electric potential is commonly interpreted as nothing more than notation for representing the electric field rather than as something real itself. However, QM makes an interesting prediction regarding a double-slit apparatus where for some distance beyond each slit, the particle paths are entirely contained in metal pipes (figure 9.4). Because the pipes are metallic, the electrons on their surfaces can freely arrange themselves to counteract any electric field outside of them, so that $\boldsymbol{E} = 0$ in any region well inside a pipe. Suppose the pipes are made electrically neutral relative to each other while a particle is traveling from the source to and through the double-slitted barrier. But once the particle arrives deep within the pipe, suppose that one pipe is given a charge relative to the other. Then $\boldsymbol{E}$ is non-zero in the region outside of the pipes and there is a potential difference between the region inside one pipe and the region inside the other. Once the particle nears the pipe exit, suppose that the charges are removed from the pipe to which they were added. So once again, $\boldsymbol{E} = 0$ everywhere and there is no potential difference between the regions inside the pipes. Throughout its journey from source to screen, then, the particle passes exclusively through regions where $E = 0$. However, the pattern of hits that accumulates on the screen is shifted from the original case where no potential difference was applied. This result, which has been verified experimentally, is a purely quantum-mechanical phenomenon; nothing like it is predicted by classical electromagnetism. (After all, the particles feel no *force* since $E = 0$ wherever they are. So in classical physics, they could not be affected!)

How are we to interpret this result (called the "electric Aharanov–Bohm effect" after the two physicists who predicted it in 1959)?[6] If E causes the particle's wave-function to change, shifting the pattern of hits, then the cause is not local to its effect, since E is non-zero only in regions through which the particle does not pass. At the particle's locations, $E = 0$ whether or not one of the pipes is charged. So the effect must be caused not by the local E, but by a distant E. On the other hand, the electric potential's value might cause the particle's wave-function to change. The electric potential of one pipe's interior relative to the other's will be different depending on whether or not the pipe is charged. Of course, for the potential to be a cause of anything, the potential (and not merely the field) must be real.

Notice that even on this interpretation, the potential's *absolute value* is irrelevant to the effect. The probability that a particle will hit a given spot on the screen depends on a potential difference: how the potentials at the locations through which a particle would pass in traveling from the *upper* slit to the spot on the screen *differ from* the potentials at the locations through which a particle would pass in traveling from the *lower* slit to the spot on the screen. This potential difference does not change if the potentials everywhere are raised or lowered by the same amount.

However, if a particle travels through only *one* slit, then since the above potential *difference* matters, the potentials at locations through which the particle does *not* pass affect the outcome. So again, we have a violation of spatial locality, even if the potential is real – unless a single particle somehow passes through *both* slits on its way from the source to the screen.

5   Entanglement, Reduction, and Intrinsic Properties

I cannot survey comprehensively the various ways in which QM's apparent nonlocality has been interpreted.[7] But I shall explore one avenue that might naturally occur to you given our discussion in chapter 1 of intrinsic properties and causal relevance. This interpretation seems to reconcile QM with spatiotemporal locality, but only in the narrowest sense. It does not eliminate the spooky connections, although it avoids interpreting them as *causal* connections – as action at a distance.

Suppose that two bodies stand in a certain relation to each other. For example, suppose that they differ in charge by 3 statcoulombs. Ordinarily, we understand this to mean that one body has some definite charge (say, 5 statcoulombs), that the other has some definite charge (say, 2 statcoulombs), and that these two charges differ by 3 statcoulombs. In other words, the two bodies stand in their relation in virtue of their *intrinsic properties*. Recall from chapter 1 that an intrinsic property is, roughly speaking, a property that a thing possesses entirely in virtue of what that thing itself is like. For something to have a certain intrinsic property here now does not logically require the cooperation of any other region of the universe or of the universe at any other moment. So, for instance, a body's having 5 statcoulombs of charge now is one of its intrinsic properties, whereas a body's now being charged to 3 statcoulombs more than another body is not one of its intrinsic properties, since the first body possesses this property partly by virtue of something about the second body.

Now as I said, two bodies stand in various relations (such as differing in charge by 3 statcoulombs) in virtue of their intrinsic properties. Apparently, given the first body's intrinsic properties along with the second's, all of their relations to each other are determined. Furthermore, corresponding to any relation in which two bodies stand to each other, there is an *intrinsic* property *of the pair* of bodies, such as consisting of two bodies that differ in charge by 3 statcoulombs. Common sense suggests that the first body's intrinsic properties, along with the second's, determine the intrinsic properties of the pair.

The only qualification to this idea that obviously must be made concerns spatiotemporal properties and relations. Although some of our spatial relations (such as your now being 5 inches taller than I) are determined by our intrinsic properties now (such as your height and mine), what about your now being 5 feet away from me? Likewise, the system of you and me possesses the intrinsic property of including two bodies 5 feet apart. Whether that fact is determined by your intrinsic properties and mine depends on whether a body's being at such-and-such location now is one of its intrinsic properties. Perhaps it is, but on the other hand, perhaps it is a relation in which the body stands to some points in space, or perhaps it is a relation in which the body stands to something else (such as a frame of reference). This is not at all obvious (see box 9.3).

However, we could try to amend our principle to cover such spatiotemporal properties and relations. Consider the first body's intrinsic properties *and* some select group of its spatiotemporal properties (whether they are intrinsic, or involve its relations to points in space-time, or involve its relations to certain reference frames, or whatever). Take these together with those of the second body. The revised principle says that we have thereby determined all of the bodies' relations to each other and all of the pair's intrinsic properties. Of course, this principle becomes more interesting as the select group of spatiotemporal properties becomes more exclusive. (For example, if

---

Box 9.3
Relativity and spatiotemporal relations

Relativity adds further complications. The relationship of two bodies' moving at speed v relative to each other is not determined by the intrinsic properties of the first body plus the intrinsic properties of the second. That is because in relativity, a body does not have an *absolute* velocity, so the bodies' relative speed is not given by the difference between their absolute velocities. Moreover, while the first and second bodies' velocities *relative* to a given frame *do* determine their relative speed, a body's velocity relative to a given frame is not obviously among its *intrinsic* properties. (It concerns the body's *relation* to a certain reference frame.)

Of course, we might take our initial principle that the intrinsic properties of a pair of bodies are determined by the first body's intrinsic properties along with the second's, and introduce an exception for the pair's intrinsic properties that express the two bodies' spatiotemporal relations (such as their relative speed). But under relativity, this exception is more costly than you might expect. For example, consider the mass of the system consisting of the two bodies. It is an intrinsic property of the system, but (recalling equation 8.8) it depends not just on the two constituents' masses, but also on their speeds in the $p = 0$ frame. So perhaps the pair's mass must qualify as an exception to the principle — in other words, as expressing (at least in part) the two bodies' spatiotemporal relations.

the group of spatiotemporal properties includes the distance of each of the two bodies from every other body, then it is uninteresting that the two bodies' being 5 feet apart is determined by the first body's intrinsic properties, the second's, and the specified group of spatiotemporal properties.)

Whatever intuition motivates our principle, that intuition applies not only to *bodies*, but to physical objects in general. Suppose the electric field here now and the electric field there now stand in some relation; perhaps they point in opposite directions, or one might be twice the magnitude of the other. The motivation behind our principle suggests that this relation is determined by E's intrinsic properties here now and E's intrinsic properties there now. Likewise, consider two physical objects that are material but perhaps not full-fledged bodies. For instance, each could be merely part of a body, such as the Earth's northern hemisphere and Mars's southern hemisphere. Their relations, too, are presumably determined by the intrinsic properties of each, along with some small subset of the spatiotemporal properties of each. It is difficult to know how far to push this principle. Applied to events, for instance, it suggests that the relations between two events are determined by the intrinsic properties of each (along with some reasonable set of their spatiotemporal properties). This seems sensible enough for many sorts of relations between events, such as one event's involving the occurrence of a force twice as great as the other event involves. But the same principle does not obviously apply to causal relations between two events. Let us set this issue aside.

Our principle says roughly that a whole, such as a particle pair, is nothing more than the "sum" of its parts – that is, the whole is "reducible" to its parts – and that each of those parts, in turn, is nothing more than the "sum" of its parts, and so on down, until we arrive at the most fundamental physical "objects" in some broad sense of the word. Accordingly, let us call this idea "the reduction principle."[8] If QM is complete, then a particle pair that we used to elaborate Bell's inequalities appears to constitute a counterexample to even a generous form of the reduction principle. When a pair of particles is emitted by the source (figure 9.1), here is one of the relations between the two particles: measurement of their x-spins would yield opposite outcomes. But this relation is not determined by the pre-measurement x-spin state of the left particle and the pre-measurement x-spin state of the right particle. That is because before the measurement is made, it is not the case that one particle in the pair has the intrinsic property of being |x+ > and the other has the intrinsic property of being |x– >. If QM is complete, neither particle is in a pure x-spin state before an x-spin measurement is made.

Let us examine this more carefully. What intrinsic properties of the left particle together with what intrinsic properties of the right particle could possibly combine to give the particle pair its disposition to produce opposite x-spin measurement outcomes on the two wings? Since the left and right particles before measurement are not in pure x-spin states, it might be supposed that the left particle's intrinsic x-spin state is the superposition

$$(\sqrt{0.5})|xL+ > - (\sqrt{0.5})|xL– >,$$

and the right particle's is

$$(\sqrt{0.5})|xR+ > - (\sqrt{0.5})|xR– >.$$

So upon being measured for its x-spin, the left particle has a 50 percent chance of returning x+ and a 50 percent chance of returning x–, and the same for the right particle. But these superpositions fail to determine the *relation* between the two particles; they do not preclude the measurements' returning the same x-spin component (xL+,xR+ or xL–,xR–), whereas in fact, they can return only opposite x-spin components (xL+,xR– or xL–,xR+).

How can this example, which is often described as involving two "entangled" particles, be reconciled with the apparently sensible (though somewhat vague) reduction principle: that a relation between two physical objects holds entirely in virtue of the two objects' own intrinsic properties (and some of their spatiotemporal properties)? Here is an avenue of escape that might cross your mind: the example does not actually involve two physical objects! Neither particle (on this view) has any intrinsic properties of its own prior to measurements being made. The particle *pair as a whole* is in a certain state, which is characterized by the pair's wave-function, the x-spin part of which is the superposition

$$(\sqrt{0.5})|xL+ > |xR– > - (\sqrt{0.5})|xL– > |xR+ >.$$

But neither of the two particles has its own wave-function, its own state; it isn't even in a superposition of pure x-spin states. Therefore, the particle pair, prior to measurement, does not really consist of two objects. (However, I will somewhat misleadingly continue to refer to it as a particle *pair*.) Obviously, the particle pair occupies two volumes separated by a region of space. (So does the system consisting of the Earth's northern hemisphere and Mars's southern hemisphere.) But it would not even be correct to say that the particle pair has two *parts*, since those "parts" lack states of their own. In contrast, a real part of a body (such as the Earth's northern hemisphere) has intrinsic and relational properties of its own. It is a "something"; it possesses various properties as a matter of objective fact, and so it exists as a metaphysically separate thing (albeit physically attached to the rest of the Earth).

Since a particle pair before measurement fails to consist of two objects, it does not violate the reduction principle, which concerns relations between two physical objects. Of course, since the particle pair is itself an object, the principle does say something about its relations to other objects. But the particle pair then complies with the reduction principle. For instance, suppose we have two objects each of which is an entangled "pair of particles" possessing the *intrinsic* property of being certain to produce opposite x-spin measurement outcomes on the two wings. Then the two objects stand in a certain relation: they have the same x-spin state. That relation is determined by the first particle pair's intrinsic x-spin state (namely, $(\sqrt{0.5})|xL+ > |xR- > - (\sqrt{0.5})|xL- > |xR+ >$) and the second's (which is the same), in accordance with the reduction principle.[9]

With this view in mind, let us think about causes and effects when the first measurement on a particle pair takes place – say, an x-spin measurement on the left wing of the apparatus. Previously, we might have said that when the measurement occurs, the wave-function of the particle on the right collapses into a pure x-spin state. Apparently, this event on the right wing has two causes: the particle pair's being in its pre-measurement state and the left measuring device's being in its initial state (namely, set to measure x-spin rather than y-spin or z-spin). The former cause is local to the effect (since the two "touch" on the right wing), but the latter cause is not local to the effect; there is a spatial gap separating them. An alternative proposal is that the state of the left wing's measuring device and the particle pair's state together locally cause the left particle's wave-function to collapse to $|x+ >$, and this collapse, in turn, instantly causes the wave-function of the right particle to collapse to $|x- >$. In the second step, there is again a region of space separating cause and effect.

Both of these proposals refer to the effect as the right particle's wave-function collapsing to a pure x-spin state. But this is incorrect, according to the view that a pre-measurement particle pair fails to consist of two objects: prior to the left measurement, the right particle had no wave-function to collapse! For that matter, neither did the left particle. Rather than separate effects occurring on the two wings, there is a single effect, an event occurring at the left measuring device and in a region on the right wing. Recall from chapter 1 that an event need not occur at a single point in space; an event can take place over an entire region, as when the shattering of a window occurs throughout the volume that the window occupied. Of course, that region was continuous, whereas I am now discussing an event that allegedly takes place in a region consisting of two volumes with a space between them. (I'll return to this point.) Nevertheless, if there weren't two particles with their own states, but only a single object (the particle pair), we must consider the effect as the collapse of their joint wave-function. This collapse occurs in a region on the left wing and a region on the right wing. Hence, spatial locality is not violated, since there is a complete set of causes (consisting of the particle pair's being in a certain pre-measurement state and the left measuring device's being in a certain pre-measurement state) no member of which is separated from the effect by a finite non-zero distance.

In other words, the effect is the "disentangling" of the particle pair; that object is becoming *two* particles, each with its own wave-function. So the effect – the disentangling – occurs over both wings of the apparatus. Admittedly, if the right particle were simply getting its x-spin state *changed* from one value (which may even be a superposition) to another, then it would be difficult to regard what happens to the right particle as an event extending over both wings. It would merely be a change in one of the right particle's intrinsic properties. But on the interpretation we are now discussing, there was no separate right particle or separate left particle before a measurement was made; neither had intrinsic properties since neither had a state of its own. There was just a single object – which, upon being measured, fell apart into two particles.

In chapter 1, we saw that the same event can be picked out in many different ways – for example, as the shattering of the window or as what made the noise that startled the neighbors. In the two-wing experiment we have been exploring, the coming into being on the left of a particle in the $|x + >$ state is the same event as the coming into being on the right of a particle in the |x – > state. Another way of picking out the event is as the collapse of the particle pair's wave-function. This event occurs on both wings of the apparatus since those are the locations where immediately after the collapse, there is some non-zero probability of finding one of the two particles into which the particle pair is transformed.

Analogously, a single electron that has just passed through the double-slit apparatus must exist in regions behind *both* slits, since its wave-function after passing through the barrier is a superposition of pure |through top slit> and pure |through bottom slit> states. This interpretation of the electron's trajectory stops cold our earlier argument that spatial locality is violated in the double-slit case. According to that argument, an electron can reach a given spot on the screen only by passing through the top slit and then traveling to that spot or by passing through the bottom slit and then traveling to the spot. These two options are mutually exclusive. Hence, the probability of a hit at a given spot on the screen, if both slits are open, equals the probability that the source will emit an electron that passes through the top slit and then travels to that spot plus the probability that the source will emit an electron that passes through the bottom slit and then travels to that spot:

$$\text{pr(hit|2)} = \text{pr(hit from bottom|2)} + \text{pr(hit from top|2)}. \quad (9.34)$$

However, on the interpretation that we are now considering, the electron does not pass through one slit *rather than* the other. So (9.34) does not follow.

This interpretation thereby also avoids a problem we encountered when we tried to interpret the electric potential as a real, local cause of the electron's behavior giving rise to the Aharanov–Bohm effect. That effect depends on the potential differences between corresponding regions behind the two slits. If the electron passes through only one slit, then the potentials at locations behind the other slit can affect it only nonlocally. But if the electron is located at all points where its wave-function gives it a non-zero probability of being detected, then the potentials behind both slits can be local causes of the electron's behavior.

But enough! We cannot close our eyes forever to the serious difficulties in this interpretation. How can an electron be completely contained in a certain region consisting of two volumes with a space between them without the electron's being definitely in one of those volumes rather than the other? This is really once again the question that Schroedinger's cat paradox asks: What the heck is a superposition? This question applies to the cat's state (a superposition of life and death), to the electron's state just after passing through the double-slit barrier (a superposition of |through top slit> and |through bottom slit>), and to the entangled pair's state (a superposition of |xL+ > |xR− > and |xL− > |xR+ >).

Without an answer to this question, this interpretation's strategy for reconciling QM with locality and the reduction principle begins to seem rather empty. The interpretation specifies that just after passing through the double-slit barrier, the electron is present in the region behind the bottom slit and in the region behind the top slit, and no event involving the electron is confined to just one of those regions. Analogous remarks apply to the entangled particle pair. Strictly speaking, then, both spatiotemporal locality and the reduction principle are satisfied. The whole particle pair isn't anything more than the sum of its parts since it doesn't have genuine "parts" on the left and right, so the reduction principle is satisfied. The wave-function collapse occurs over both wings because there aren't separate physical objects on the left and right until after the measurement has taken place, so locality is satisfied. But how can it be that there aren't full-fledged physical objects on the left and right wings? That there aren't may follow from the fact that the system's wave-function cannot be decomposed into separate left and right components. But this does not bring us much closer to understanding what's really going on.

In other words, spatiotemporal locality as defined in chapter 1 is satisfied under this interpretation, but that is only because the apparent "nonlocality" has gone underground: instead of involving causal relations (which are addressed by the locality principles we defined in chapter 1), this interpretation places the apparent "nonlocality" at a more basic level. The "nonlocality" enters at the level of the events standing in causal relations. An event occurs on the two wings, in regions separated by a spatial gap. But this event cannot be decomposed into an event taking place on the left wing and another event taking place on the right wing. This is weird. However, because the weirdness pertains to the character of the events themselves rather than to their causal relations, spatiotemporal locality is satisfied: this interpretation avoids hypothesizing any action at a distance.

Here is another way to appreciate that this interpretation upholds locality only in a very thin sense. On this interpretation, a complete set of causes of the collapse of the particle pair's wave-function consists of the particle pair's pre-measurement wave-function and the left measuring device's pre-measurement state. Thus, there is a location at which the effect takes place (namely, on the right wing) and a cause in the set (namely, the left measuring device's being in a certain pre-measurement state) such that this effect-location is separated by some region from the nearest location at which this cause takes place. This by itself is not suspicious: if the window's shattering is caused by a baseball's colliding with it, then if the baseball actually comes in contact with only the *center* of the window, then a region somewhere along the window's periphery is an effect-location that is separated by some gap from the nearest location at which the cause takes place. However, between the window's center and its periphery, there was continuous window, throughout which shattering took place. The same cannot be said of the particle pair. Every continuous path from the cause-location on the left wing to the effect-location on the right wing passes through some points where the effect does not take place. (And there is no complete set of the effect's causes that lacks this feature.) This certainly seems like action at a distance! But it gets papered over by the fact that the effect, on this interpretation, is not confined to the right wing. Other locations at which the effect occurs (namely, on the left wing) coincide with locations where this cause occurs. So spatial locality is satisfied.

Locality, as defined in chapter 1, fails to require that an event serving as a cause or effect occupy a continuous region in space. For example, locality does not prohibit an effect (such as the collapse of the particle pair's wave-function) from occupying two volumes with a space between them. If some cause (such as the left measuring device's having a certain pre-measurement state) is confined to one of those volumes, then the effect's occurrence in the other volume will have something of the weirdness of action at a distance. But, strictly speaking, it will not count as violating locality.

Perhaps what we have just learned is that the locality principles defined in chapter 1 capture our intuitive notion of locality – of action by contact, as contrasted with action at a distance – only when those principles have been supplemented by certain assumptions. One assumption is that every event serving as a cause or effect occupies a spatiotemporally continuous region. Another assumption is that in any spatiotemporally continuous part of the region occupied by an effect, even a part as small as a point in space at a moment in time, there is an effect occurring throughout that part but nowhere beyond it. So in order for the reduction principle and spatiotemporal locality to hold, they must apply to every part of the original effect. These assumptions are commonsensical, but as we have seen, QM may perhaps best be interpreted as denying them. In that case, QM does not *violate* locality – does not include action at a distance – but rather violates something presupposed by our intuitive distinction between action by contact and action at a distance.

Einstein seems to have had assumptions like these in mind when he wrote:

> [W]hatever we regard as existing (real) should somehow be localised in time and space. That is, the real in part of space A should (in theory) somehow "exist" independently of what is thought of as real in space B. When a system in physics extends over the parts of space A *and* B, then that which exists in B should somehow exist independently of that which exists in A. That which really exists in B should therefore not depend on what kind of measurement is carried out in part of space A; it should also be independent of whether or not any measurement at all is carried out in space A. If one adheres to this programme, [then one must believe that the left and right particles are characterized by hidden variables, since QM fails to ascribe to them states of their own, and so] one can hardly consider the quantum-theoretical description as a complete representation of the physically real. If one tries to do so in spite of this [in other words, if one regards QM as complete], one has to assume that the physically real in B suffers a sudden change as a result of a measurement in A. [Spooky action at a distance!] My instinct for physics bristles at this. However, if one abandons the assumption that what exists in different parts of space has its own, independent, real existence, then I simply cannot see what it is that physics is meant to describe. For what is thought to be a "system" is, after all, just a convention, and I cannot see how one could divide the world object-ively in such a way that one could make statements about parts of it. (Born 1971: 164–5)

Clearly, in insisting that the entangled particle pair cannot be re-garded as two objects separated by a region of space, and likewise that the collapse of the particle pair's wave-function does not consist of one event on the left wing and another on the right, the interpretation we have been examining has "abandon[ed] the assumption that what exists in different parts of space has its own, independent, real existence. . . ."

In relativity (as we saw in chapter 8), we can regard the fragments produced when a radioactive nucleus decays either as separate objects (with their own masses) or as forming a single object (with a mass generally unequal to the sum of the separate objects' masses). Which "perspective" we select for dividing the universe into individual objects is, in Einstein's words, "just a convention"; neither perspective is more accurate than the other. I used this idea to argue that the "conversion" of mass into energy (or vice versa) is not a real physical process. However, under the interpretation of QM that we have been looking at, it is a matter of fact rather than convention that the particle pair is a single object and not two separate objects.

Why should this interpretation make it difficult to "see what it is that physics is meant to describe"? Because to describe the universe, physics must describe the objects in it: physics must separate the universe into things (whether bodies, fields, or whatever), specify the kinds of properties that those things can possess (and can perhaps be measured to possess), and identify the laws governing the changes and interactions of those things. But virtually every time two particles interact and are not measured, they become entangled. If they then have no separate existence, it is hard to see what the objects described by physical theory can be. Of course, we could still use QM to predict the probabilities of various measurement outcomes. But we would be unable to interpret our theories as describing anything less than the entire universe – or, at least, very large parts of it. They alone would qualify as physical objects.

In particular, charged particles in any two regions of space-time, between which the spatiotemporal interval $I > 0$, are interacting gravitationally and electromagnetically. Likewise (as we saw in chapter 8), electromagnetic fields in any two such regions have mass and so are interacting gravitationally. Therefore (unless elaborate measurements are being made to collapse the wave-function), QM must describe the two regions as a single object (Howard 1989: 240–53); they are entangled, like the particles on the left and right wings of the apparatus in figure 9.1. (However, the correlations between the regions may be negligible for all practical purposes.)

None of these remarks should be understood as an argument that this interpretation of QM is false. Rather, they merely emphasize how much further we have to go in developing an adequate conceptual scheme for interpreting quantum physics. This project remains the focus of intense work among natural philosophers today.

Discussion Questions

You might think about . . .

1 Suppose a single unentangled particle is fired from a source toward a screen equipped with a detector that will indicate where on the screen the particle arrives. QM determines the probabilities of detection at various spots on the screen but does not determine precisely where the particle will arrive. When the particle is detected at a certain spot on the screen, its wave-function instantly changes at various other locations on the screen. Since these changes are caused by the particle's being detected some distance away, is spatial locality being violated? (See Einstein 1928.)

2 Explain the differences among the following interpretations of Heisenberg's "uncertainty principle":

(a) It is impossible to know, with arbitrarily great precision, the x-spin and y-spin of a given particle at a given moment.
(b) A particle does not simultaneously possess a definite x-spin and a definite y-spin.
(c) If we measure a particle's x-spin, we inevitably and uncontrollably (that is, randomly) disturb the particle's y-spin to some degree. The disturbance increases in proportion to the precision of our measurement.
(d) QM simply forbids our asking whether or not a particle simultaneously possesses a definite x-spin and a definite y-spin.

3 Is a particle's wave-function a dispositional property? (Recall chapter 3.)

4 Consider these three claims:

(a) The property of having a spin that is entangled with another particle's (in other words, the property of being "spin-entangled") is not an intrinsic property.
(b) Suppose the fact that an object possesses property P logically guarantees that it possesses property Q (in the way that the fact that something is square guarantees that it has four sides). Then if P is an intrinsic property, so is Q.

(c)   P is an intrinsic property if and only if the property of not being
      P is an intrinsic property.

Claim (a) seems plausible: for a particle (say, the particle on the left
wing of the apparatus) to be spin-entangled, there must somewhere be
another particle with which it is entangled; it cannot qualify as entangled solely in virtue of whatever properties it has intrinsically.[10] Claim
(b) seems plausible: if being Q depends on the cooperation of the rest
of the universe, and being P requires being Q, then being P depends
on the universe's cooperation. Finally, claim (c) seems plausible, as I
explained in chapter 1. Now show that these three claims lead to the
very implausible conclusion that the property of having some pure
(and hence unentangled) spin state, such as $|x+>$, is *not* intrinsic.
Accordingly, at least one of the three plausible-looking claims with
which I began must be mistaken. Which one(s)? Hint: Claim (b) is
false, but the flaw in it that I know about does not seem to play any
part in leading to the very implausible result of the above argument. Is
the best way to block this argument, then, to deny (a) on the grounds
that it presupposes a pair of entangled particles to consist of two objects,
each with its own intrinsic and non-intrinsic properties?

5   In section 4, I insisted that the measuring event on the left wing
apparently causes something to happen on the right wing (namely, the
right particle's acquiring some definite spin component) even though
the probability of a right measurement's having a given outcome is
unchanged by our making (say) an x-spin measurement on the left.
(In our example, the probability that xR+ is 50 percent both before
and after an x-spin measurement is made on the left.) Some have
disagreed. Here is an especially pungent remark along these lines:

> Actual calculations show that even though quantum theory is connected
> non-locally inside, these connections never get out to . . . the only as-
> pect of quantum theory that can be put to direct test. Any measurable
> influence, in terms of transmission of information, still travels at the
> speed of light or slower. . . . The perfect locality of all quantum predic-
> tions suggests that non-local connections are a theoretical artifact with
> no more reality than the dotted lines that outline the constellations on
> star maps.   (Herbert 1986: 42)

Do you agree?

Notes

1   Some interpretations of QM disagree – David Bohm's, for instance. See Albert (1992).

2   Another way around this loophole is to use an arrangement of three particles rather than two, according to a recent discovery by Daniel Greenberger, Michael Horne, and Anton Zeilinger (Mermin 1990). Bell's inequalities concern the *probabilities* that various measurements will be made. These probabilities can be compared with the frequencies with which various results are actually found in a long run of experiments, which, as the runs lengthen (and under various other assumptions, as I just mentioned), become arbitrarily likely to match the actual probabilities. But with three particles, there is an experimental arrangement in which (according to a Bell-type argument) a certain combination of measurements at the three detectors has a considerable probability of being made *in a single run of the experiment*, but that combination of measurements is completely ruled out by QM. This experiment has not been performed – yet.

3   Some philosophers, such as Price (1996), have interpreted QM as involving "backwards causation."

4   This terminology is Shimony's. A similar assumption has (confusingly for us) sometimes been called "locality" (by Jon Jarrett).

5   Such is the conventional wisdom, at any rate. From time to time, reputable physics journals carry articles proposing schemes for such communication. See, for instance, the controversy reported by Gleick (1987).

6   There is also a "magnetic Aharanov–Bohm effect," which I shall not discuss (since it involves the magnetic vector potential, which I have not introduced), but which is roughly analogous.

7   Among the best recent books on quantum metaphysics are Albert (1992), Bub (1997), Cushing and McMullin (1989), Fine (1986), Healey (1989), Hughes (1989), Maudlin (1994), Redhead (1987), and van Fraassen (1991).

8   No doubt the details of this principle require further spelling out. For instance, the intrinsic properties of the whole that are determined by the properties of the parts should presumably not include properties like "being (identical to) the Eiffel Tower" or "not being the selfsame particle as the particle that was located at such-and-such spatiotemporal location." These are sometimes (confusingly) called "qualitative" properties. Furthermore, I have not specified the precise sense in which the properties of the parts "determine" the properties of the whole. Must the properties of the parts suffice to *logically entail* the properties of the whole? Or can the properties of the parts be supplemented by the laws of nature (or some other important class of truths)?

9   Teller (1989) advances a similar line of thought, as does Richard Miller (1987: 556 and 593–4). However, my spin on this interpretation is different from Teller's. He refers to two entangled particles as exhibiting "relational holism" in that their relations fail to be determined by the intrinsic properties of the two particles so related. On the interpretation I've just floated, there aren't relations among these particles; rather, the entangled particles form a single object. The reduction principle is similar to what Teller calls "particularism." It is also similar to various principles that have been termed "separability" (e.g., Howard 1989: 226; Healey 1999: 441) or (yet again!) "locality" (Belot 1998: 540).

10   Butterfield (1993: 462) refers to "the extrinsic property of being a part of a composite."

Final Exam

Part One

Here are passages from two current physics texts. First, from Ronold W. P. King and Sheila Prasad, *Fundamental Electromagnetic Theory and Applications* (Englewood Cliffs, NJ: Prentice-Hall, 1986), p. 552:

> The fundamental law of macroscopic electromagnetism as expressed in the field and force equations is interpreted as a retarded action at a distance. The [terms] appearing in this law [such as the strength of the electric field at a given spatio-temporal location, the strength of the magnetic field there, etc.] are, of course, not assigned a *localized* physical significance as properties of a medium. The electromagnetic field and the fields of the potential functions serve merely as intermediate steps in a mathematical calculation of action between statistical distributions of charge and current.

Now from Hans Ohanian, *Classical Electrodynamics* (Boston, MA: Allyn and Bacon, 1988), p. 45:

> The electric field is a useful concept because it lets us think of electric effects as being due to local motion, or action-by-contact. In other words, we can think of the force on a charge as due to the altered state of the space surrounding the charge, not as due to action-at-a-distance of the other charge. Action-at-a-distance is unacceptable because it does not agree with relativity: if a charge could act directly on a second charge, then any change in the position of the former would be felt *immediately* at the latter, and this would make it possible to propagate

signals with a speed exceeding that of light. If the action between charges is mediated by fields, then there is no such instantaneous propagation. Any change in the position of the first charge produces a disturbance in its nearby field; and this disturbance spreads out, as a wave moving at the speed of light, from the first charge to the second.

As we will see later, fields have energy. They therefore are a form of matter; they can be regarded as the fifth state of matter (solid, liquid, gas, and plasma are the other four states of matter).

Concerning these remarks:

1   Explain *precisely* what these texts are disagreeing about. How can they *possibly be* disagreeing when they are teaching the *same* scientific theory?

2   What do King and Prasad mean by the field's and potential's serving "merely as intermediate steps in a mathematical calculation"? Is this the only alternative to interpreting them as real? On which interpretations would fields be local causes of forces?

3   Explain Ohanian's argument for the view that "action-at-a-distance is unacceptable." How would King and Prasad reply to Ohanian's argument?

4   How could Ohanian use the classical conservation laws to reply to King and Prasad? What, in classical physics, would be his reason for claiming that "fields have energy"?

5   In reply, what ontological status would King and Prasad ascribe to energy? To what extent can this issue be resolved within classical physics?

6   How does relativity theory affect the argument that fields are real by virtue of possessing energy? Does relativity theory enable Ohanian to conclude that fields are "a form of matter"?

Part Two

1   What difficulties arise in trying to understand the collision of two hard bodies?

2   How can these difficulties be avoided if the forces arising from collisions are caused by fields? Did other philosophical considerations also motivate the development of field theory?

3   Are the Earth and the Sun currently "colliding" (touching) on Faraday's view? How did Faraday argue that bodies should not be understood as made of matter – a kind of hard stuff? How strong is this argument?

4   Why, on relativity, is a body's mass not the amount of matter it contains?

Part Three

1   What is the argument from "surplus structure" against the electric potential's reality and also against energy's reality?

2   What features does this argument share with Einstein's asymmetry argument and with the argument favoring Galilean space-time over Newton's own interpretation of his mechanics?

3   What is the argument that some facts are brute, and how does this possibility affect the argument from surplus structure?

4   How strong is the argument from surplus structure? What does the discovery of the Aharanov–Bohm effect (or, as another example, the discovery of $S = Gc^2$) reveal about the strength of such arguments? What does this episode tell us about the nature of the distinction between a physical theory and its philosophical interpretation?

5   When can an argument from surplus structure be strengthened as an argument from "fine-tuning"? (Consider Einstein's asymmetry argument, for example.)

6   Why, according to Einstein, were hidden variables not mere "surplus structure" in connection with QM? How does Bell's argument affect Einstein's?

References

Abraham, Max (1951) *Classical Theory of Electricity and Magnetism*, rev. Richard Becker, trans. John Dougall, 2nd edn. New York: Hafner.

Albert, David (1992) *Quantum Mechanics and Experience*. Cambridge, MA: Harvard University Press.

Alexander, H. G. (1956) *The Leibniz–Clarke Correspondence*. Manchester: Manchester University Press.

Armstrong, D. M. (1968) *A Materialist Theory of the Mind*. London: Routledge.

Armstrong, D. M., Martin, C. B., and Place, U. T. (1996) *Dispositions: A Debate*. New York: Routledge.

Aspect, Alain, Dalibard, Jean, and Roger, Gerard (1981) Experimental tests of Bell's inequalities using time-varying analyzers. *Physical Review Letters* 49, 1804–7.

Becker, Richard (1964) *Electromagnetic Fields and Interactions* – volume 1: *Electromagnetic Theory and Relativity*, ed. Fritz Sauter. New York: Blaisdell.

Bell, John (1971) Introduction to the hidden-variable question. In B. d'Espagnat (ed.), *Foundations of Quantum Mechanics*, "Enrico Fermi" Course 49. New York: Academic Press, pp. 171–81.

Belot, Gordon (1998) Understanding electromagnetism. *British Journal for the Philosophy of Science* 49, 531–55.

Bennett, Jonathan (1988) *Events and their Names*. Oxford: Clarendon Press.

Bird, Alexander (1998) Dispositions and antidotes. *Philosophical Quarterly* 48, 227–34.

Birkeland, K. (1894) Ueber die Strahlung elektromagnetischer Energie im Raume. *Annalen der Physik* 52, 357–80.

Blackburn, Simon (1990) Filling in space. *Analysis* 50, 62–5.

Bleaney, B. I., and Bleaney, B. (1976) *Electricity and Magnetism*, 3rd edn. Oxford: Oxford University Press.

Bohm, David (1996) *The Special Theory of Relativity*. London: Routledge.

Bolzano, Bernard (1950) *Paradoxes of the Infinite*, trans. D. A. Steele. London: Routledge.

Born, Max (1924) *Einstein's Theory of Relativity*. London: Methuen.

———(1971) *The Born–Einstein Letters*, trans. Irene Born. New York: Walker.

Born, Max, and Heisenberg, Werner (1928) La Mécanique des quanta. In *Electrons et Photons: Proceedings of the Fifth Solvay Conference*. Paris: Gauthier-Villars, pp. 143–81.

Boyer, Timothy (1971) Energy and momentum in electromagnetic field for charged particles moving with constant velocities. *American Journal of Physics* 39, 257–70.

Boyle, Robert (1991) An essay, containing a requisite digression, concerning those that would exclude the Deity from intermeddling with matter. In M. A. Stewart (ed.), *Selected Philosophical Papers of Robert Boyle*. Indianapolis: Hackett, pp. 155–75.

Brentano, Franz (1988) *Philosophical Investigations on Space, Time, and the Continuum*, trans. B. Smith. London: Croom Helm.

Bridgman, Percy (1941) *The Nature of Thermodynamics*. Cambridge, MA: Harvard University Press.

———(1958) *The Logic of Modern Physics*. New York: Macmillan.

Broad, C. D. (1933) *Examination of McTaggart's Philosophy*. Cambridge, UK: Cambridge University Press.

Bub, Jeffrey (1997) *Interpreting the Quantum World*. New York: Cambridge University Press.

Buchwald, Jed (1985) *From Maxwell to Microphysics*. Chicago: University of Chicago Press.

Butterfield, Jeremy (1993) Interpretation and identity in quantum theory. *Studies in the History and Philosophy of Science* 24, 443–76.

Calkin, M. G. (1966) Linear momentum of quasistatic electromagnetic fields. *American Journal of Physics* 34, 921–5.

Campbell, Norman (1957) *Foundations of Science*. New York: Dover.

Cao, Tian Yu (1997) *Conceptual Developments of 20th Century Field Theories*. Cambridge, UK: Cambridge University Press.

Carter, G. W. (1954) *The Electromagnetic Field in its Engineering Aspects*. London: Longman.

Cartwright, Helen Morris (1975) Amounts and measures of amount. *Nous* 9, 143–64.

Cat, Jordi (1998) Maxwell's problem of understanding potentials concretely. Preprint #101, Max Planck Institute for the History of Science.

Clauser, John, and Horne, Michael (1974) Experimental consequences of objective local theories. *Physical Review D* 10, 526–35.

Clauser, John, and Shimony, Abner (1978) Bell's theorem: experimental tests and implications. *Reports on Progress in Physics* 41, 1881–1927.

Cohen, I. B. (ed.) (1978) *Isaac Newton's Papers and Letters on Natural Philosophy*. Cambridge, MA: Harvard University Press.

Cole, Daniel (1999) Cross-term conservation relationships for electromagnetic energy, linear momentum, and angular momentum. *Foundations of Physics* 29, 1673–85.

Collins, C. B., and Hawking, Stephen (1973) Why is the universe isotropic? *Astrophysical Journal* 180, 317–34.

Comte, Auguste (1968) *Cours de la philosophie positive*. Paris: Anthropos.

Corson, Dale, and Lorrain, Paul (1962) *Introduction to Electromagnetic Fields and Waves*. San Francisco: Freeman.

Cotts, Robert, and Detenbeck, Robert (1966) *Matter in Motion*. Seattle: University of Washington Press.

Cushing, James, and McMullin, Ernan (eds.) (1989) *Philosophical Consequences of Quantum Theory*. Notre Dame, IN: University of Notre Dame Press.

Darrigol, Oliver (1996) The electrodynamic origins of relativity theory. *Historical Studies in the Physical and Biological Sciences* 26, 241–312.

Davidson, Donald (1980) Causal relations. In *Essays on Actions and Events*, Oxford: Clarendon Press, pp. 149–62.

Davies, P. C. W., and Brown, J. R. (eds.) (1986) *The Ghost in the Atom*. Cambridge, UK: Cambridge University Press.

Dennett, Daniel (1987) *The Intentional Stance*. Cambridge, MA: MIT Press.

——— (1991) Real patterns. *Journal of Philosophy* 88, 27–51.

d'Espagnat, Bernard (1979) The quantum theory and reality. *Scientific American* 241 (5), 158–81. (European edition, 128–40.)

Dirac, P. A. M. (1938) Classical theory of radiating electrons. *Proceedings of the Royal Society Series A* 167, 149–69.

Djuric, Jovan (1975) Spinning magnetic fields. *Journal of Applied Physics* 46, 679–88.

——— (1979) Reply to "Comment on 'Spinning magnetic fields.'" *Journal of Applied Physics* 50, 537.

Earman, John (1987) Locality, nonlocality, and action at a distance. In R. Kargon and P. Achinstein (eds.), *Kelvin's Baltimore Lectures and Modern Theoretical Physics*. Cambridge, MA: MIT Press, pp. 449–90.

——— (1989) *World Enough and Space-Time*. Cambridge, MA: MIT Press.

Ehring, Douglas (1999) Tropeless in Seattle: the cure for insomnia. *Analysis* 59, 19–24.

Einstein, Albert (1928) [Comment]. In *Electrons et Photons: Proceedings of the Fifth Solvay Conference*. Paris: Gauthier-Villars, p. 255.

——— (1931) *Relativity: The Special and the General Theory*. New York: Crown.

——— (1934) *Essays in Science*. New York: Philosophical Library.

——— (1935) Elementary derivation of the equivalence of mass and energy. *Bulletin of the American Mathematical Society* 41, 223–30.

—— (1944) Remarks on Bertrand Russell's theory of knowledge. In Paul Arthur Schilpp (ed.), *The Philosophy of Bertrand Russell*. Evanston, IL: Northwestern University Press, pp. 279–91.

—— (1949) Autobiographical notes. In Paul Arthur Schilpp (ed.), *Albert Einstein: Philosopher-Scientist*. New York: Tudor, pp. 1–95.

—— (1952) The foundation of the general theory of relativity. In W. Perrett and G. B. Jeffery (eds.), *The Principle of Relativity*. New York: Dover, pp. 109–64.

—— (1953) *The Meaning of Relativity*. Princeton, NJ: Princeton University Press.

—— (1989) *The Collected Papers of Albert Einstein*, vol. 2, trans. Anna Beck. Princeton, NJ: Princeton University Press.

Elkana, Yuhuda (1974) *The Discovery of the Conservation of Energy*. Cambridge, MA: Harvard University Press.

Eyges, Leonard (1972) *The Classical Electromagnetic Field*. New York: Dover.

Faraday, Michael (1839) *Experimental Researches in Electricity*, vol. 1. London: Richard and John Edward Taylor.

—— (1844) A speculation touching electric conduction and the nature of matter. In *Experimental Researches in Electricity*, vol. 2. London: Richard and John Edward Taylor, pp. 284–93.

—— (1852) Experimental researches in electricity – twenty-eight series. *Philosophical Transactions of the Royal Society* 142, 25–56.

—— (1855) On atmospheric magnetism. In *Experimental Researches in Electricity*, vol. 3. London: R. Taylor and W. Francis, pp. 323–7.

—— (1935) *Faraday's Diary*. London: G. Bell and Sons.

Feynman, Richard (1965) *The Character of Physical Law*. Cambridge, MA: MIT Press.

Feynman, Richard, Leighton, Robert, and Sands, Matthew (1963) *The Feynman Lectures on Physics*. Reading, MA: Addison-Wesley.

Fine, Arthur (1977) Appendix. *Journal of Philosophy* 74, 538.

—— (1986) *The Shaky Game*. Chicago: University of Chicago Press.

—— (1991) Inequalities for nonideal correlation experiments. *Foundations of Physics* 21, 365–78.

FitzGerald, George (1895) The foundations of dynamics. *Nature* 51, 283–5.

Foster, John (1982) *The Case for Idealism*. London: Routledge.

Friedman, Michael (1974) Explanation and scientific understanding. *Journal of Philosophy* 71, 5–19.

Furry, W. H. (1969) Examples of momentum distributions in the electromagnetic field and in matter. *American Journal of Physics* 37, 621–36.

Geroch, Robert (1978) *General Relativity: From A to B*. Chicago: University of Chicago Press.

Gleik, James (1987) In defiance of Einstein, physicists seek faster-than-light messages. *New York Times* (October 27), 16.

Glymour, Clark, Scheines, Robert, Spirtes, Peter, and Kelly, Kevin (1987) *Discovering Causal Structure*. New York: Academic Press.

Good, John Mason (1837) *The Book of Nature*. Hartford, CN: Belknap Press.

Gooding, David (1978) Conceptual and experimental bases of Faraday's denial of electrostatic action at a distance. *Studies in History and Philosophy of Science* 9, 117–49.

——— (1980) Faraday, Thomson, and the concept of the magnetic field. *British Journal for the History of Science* 13, 91–120.

Goodman, Nelson (1983) *Fact, Fiction, and Forecast*, 4th edn. Cambridge, MA: Harvard University Press.

Graham, G. M. and Lahoz, D. G. (1980) Observation of static electromagnetic angular momentum *in vacuo*. *Nature* 285, 154–5.

Halliday, David, Resnick, Robert, and Krane, Kenneth (1992) *Physics*, 4th edn., extended. New York: Wiley.

Harman, Peter M. (1985) Edinburgh philosophy and Cambridge physics. In Harman (ed.), *Wranglers and Physicists*. Manchester: Manchester University Press, pp. 202–24.

——— (1998) *The Natural Philosophy of James Clerk Maxwell*. Cambridge, UK: Cambridge University Press.

Harre, Rom, and Madden, E. H. (1975) *Causal Powers*. Totowa: Rowman and Littlefield.

Healey, Richard (1989) *The Philosophy of Quantum Mechanics: An Interactive Interpretation*. Cambridge, UK. Cambridge University Press.

——— (1999) Quantum analogies: a reply to Maudlin. *Philosophy of Science* 66, 440–7.

Heaviside, Oliver (1922) *Electromagnetic Theory*. London: Benn.

——— (1925) *Electrical Papers*. Boston, MA: Copley.

Heilbron J. L. (1982) *Elements of Early Modern Physics*. Berkeley: University of California Press.

Heisenberg, Werner (1958) *Physics and Philosophy*. London: George Allen and Unwin.

Hempel, C. G. (1973) Science unlimited? *Annals of the Japan Association for the Philosophy of Science* 4, 181–202.

——— (1978) Dispositional explanation. In Raimo Tuomela (ed.), *Dispositions*. Dordrecht: Reidel, pp. 137–46.

Herbert, Nick (1986) How to be in two places at one time. *New Scientist* (August 21), 41–4.

Hertz, Heinrich (1893) *Electric Waves*, trans. D. E. Jones. London: Macmillan.

——— (1994) Introduction to *Principles of Mechanics*. In Joseph F. Mulligan (ed.), *Hertz: A Collection of Articles and Addresses*. New York: Garland, pp. 323–64.

Hesse, Mary (1965) *Forces and Fields*. Totowa: Littlefield.

Hirosige, Tetu (1976) The ether problem, the mechanistic worldview, and the origins of the theory of relativity. *Historical Studies in the Physical Sciences* 7, 7–82.

Hoefer, Carl (2000) Energy conservation in GTR. *Studies in History and Philosophy of Modern Physics* 31B, 187–99.

Holton, Gerald (1973) *Thematic Origins of Scientific Thought*. Cambridge, MA: Harvard University Press.

Howard, Don (1989) Holism, separability, and the metaphysical implications of the Bell experiments. In J. Cushing and E. McMullin (eds.), *Reflections on Bell's Theorem*. Notre Dame, IN: University of Notre Dame Press, pp. 224–71.

Hughes, R. I. G. (1989) *The Structure and Interpretation of Quantum Mechanics*. Cambridge, MA: Harvard University Press.

Hume, David (1977) *An Enquiry Concerning Human Understanding*. Indianapolis: Hackett.

———— (1978) *A Treatise of Human Nature*, ed. L. A. Selby-Bigge, 2nd edn., rev. P. H. Nidditch. Oxford: Clarendon Press.

Hutchison, Keith (1991) Dormitive virtues, scholastic qualities, and the new philosophies. *History of Science* 29, 245–78.

Jammer, Max (1961) *Concepts of Mass*. Cambridge, MA: Harvard University Press.

Jardine, Nicholas (1984) *The Birth of History and Philosophy of Science: Kepler's "A Defense of Tycho Against Ursus."* Cambridge, UK: Cambridge University Press.

Jeans, James (1932) *The Mysterious Universe*. New York: Macmillan.

———— (1960) *The Mathematical Theory of Electricity and Magnetism*, 5th edn. Cambridge, UK: Cambridge University Press.

Jeffries, Clark (1992) A new conservation law for classical electrodynamics. *SIAM Review* 34, 386–405.

Jungnickel, Christa, and McCormmach, Russell (1986) *Intellectual Mastery of Nature*. Chicago: University of Chicago Press.

Kane, Joseph W., and Sternheim, Morton M. (1978) *Physics*. New York: Wiley.

Kant, Immanuel (1985) *Metaphysical Foundations of Natural Science*. Indianapolis: Hackett.

Kepler, Johannes (1981) *Mysterium Cosmographicum*, trans. A. M. Duncan. New York: Abaris.

Kim, Jaegwon (1974) Noncausal connections. *Nous* 8, 41–52.

Kitcher, Philip (1989) Explanatory unification and the causal structure of the world. In P. Kitcher and W. Salmon (eds.), *Scientific Explanation*, Minnesota Studies in the Philosophy of Science, vol. 13. Minneapolis: University of Minnesota Press, pp. 410–505.

Kline, A. David, and Matheson, Carl (1987) The logical impossibility of collision. *Philosophy* 62, 509–16.

Kneale, William (1949) *Probability and Induction*. Oxford: Clarendon Press.

Knudsen, Ole (1985) Mathematics and physical reality in William Thomson's electromagnetic theory. In P. Harman (ed.), *Wranglers and Physicists*. Manchester: Manchester University Press, pp. 149–79.

———— (1995) Electromagnetic energy and the early history of the energy principle. In A. J. Kox and D. M. Siegel (eds.), *No Truth Except in the Details*, Boston Studies in the Philosophy of Science, vol. 167. Dordrecht: Kluwer, pp. 55–78.

Konopinski, E. J. (1978) What the electromagnetic vector potential describes. *American Journal of Physics* 46, 499–502.

Lange, Marc (2000) *Natural Laws in Scientific Practice*. New York: Oxford University Press.

Langton, Rae (1998) *Kantian Humility*. New York: Oxford University Press.

Larmor, Joseph (1897) A dynamical theory of the electric and luminiferous medium – Part III. *Philosophical Transactions of the Royal Society Series A* 190, 205–300.

Leslie, John (1989) *Universes*. London: Routledge.

Levere, Trevor (1968) Faraday, matter, and natural theology – reflections on an unpublished manuscript. *British Journal for the History of Science* 4, 95–107.

———— (1971) *Affinity and Matter*. Oxford: Clarendon Press.

Lewis, David (1986) *Philosophical Papers*, vol. 2. New York: Oxford University Press.

Lewis, David, and Langton, Rae (1998) Defining intrinsic. *Philosophy and Phenomenological Research* 58, 333–45.

Lightman, Alan, and Brawer, R. (1990) *Origins*. Cambridge, MA: Harvard University Press.

Locke, John (1975) *An Essay Concerning Human Understanding*, ed. P. H. Nidditch. Oxford: Clarendon Press.

Lodge, Oliver (1885) On the identity of energy. *Philosophical Magazine* 19, 482–7.

McGuire, J. E. (1968) Force, active principles, and Newton's invisible realm. *Ambix* 15, 154–208.

Mackay, Alan (1977) *Scientific Quotations*. New York: Institute of Physics.

Mackie, J. L. (1973) *Truth, Probability, and Paradox*. Oxford: Clarendon Press.

Martin, C. B. (1994) Dispositions and conditionals. *Philosophical Quarterly* 44, 1–8.

Maudlin, Tim (1994) *Quantum Non-Locality and Relativity*. Malden, MA: Blackwell.

———— (1996) On the unification of physics. *Journal of Philosophy* 93, 129–44.

Maxwell, James Clerk (1879) Potential. *Electrician* 2, 271–2.

———— (1881) *An Elementary Treatise on Electricity*, ed. W. Garnett. Oxford: Clarendon Press.

———— (1890) *The Scientific Papers of James Clerk Maxwell*, ed. W. D. Niven. Cambridge, UK: Cambridge University Press.

———— (1892) *Matter and Motion*. New York: Van Nostrand.

———— (1908) *Theory of Heat*. New York: Longman.

———— (1954) *A Treatise on Electricity and Magnetism*. New York: Dover.

———— (1990) *The Scientific Letters and Papers of James Clerk Maxwell*, vol. 1, ed. P. M. Harman. Cambridge, UK: Cambridge University Press.

———— (1995) *The Scientific Letters and Papers of James Clerk Maxwell*, vol. 2, ed. P. M. Harman. Cambridge, UK: Cambridge University Press.

Mellor, D. H. (1974) In defense of dispositions. *Philosophical Review* 83, 157–81.

Mermin, N. David (1990) Quantum mysteries revisited. *American Journal of Physics* 58, 731–74.

Meyerson, Emile (1962) *Identity and Reality*, trans. Kate Loewenberg. New York: Dover.

Miller, Arthur I. (1982) On Einstein's invention of special relativity. In P. D. Asquith and T. Nickles (eds.), *PSA 1982*, vol. 2, pp. 377–402.

———— (1998) *Albert Einstein's Special Theory of Relativity*. New York: Springer.

Miller, Richard (1987) *Fact and Method*. Princeton, NJ: Princeton University Press.

Minkowski, Hermann (1952) Space and time. In W. Perrett and G. B. Jeffery, *The Principle of Relativity*. New York: Dover, pp. 73–91.

Morrison, Margaret (2000) *Unifying Scientific Theories*. New York: Cambridge University Press.

Moyer, Donald (1978) Continuum mechanics and field theory: Thomson and Maxwell. *Studies in History and Philosophy of Science* 9, 35–50.

Mumford, Stephen (1998) *Dispositions*. New York: Oxford University Press.

Newall, H. F. (1910) 1885–1894. In *A History of the Cavendish Laboratory, 1871–1910*. London: Longman, pp. 102–58.

Newton, Isaac (1999) *The* Principia: *Mathematical Principles of Natural Philosophy*, trans. I. B. Cohen and A. Whitman. Berkeley: University of California Press.

———— (1952) *Opticks*, 4th edn. New York: Dover.

Norton, John (1991) Thought experiments in Einstein's work. In G. Massey and T. Horowitz (eds), *Thought Experiments in Science and Philosophy*. Savage: Rowman and Littlefield.

Nozick, Robert (1981) *Philosophical Explanations*. Cambridge, MA: Belknap Press.

Ohanian, Hans (1976) *Gravitation and Spacetime*. New York: W.W. Norton.

Okun, Lev (1989) The concept of mass. *Physics Today* 42 (6), 31–6.

O'Railly, Alfred (1965) *Electromagnetic Theory*. New York: Dover.

Page, Leigh, and Adams, Norman (1945) Action and reaction between moving charges. *American Journal of Physics* 13, 141–7.

Pagels, Hans (1982) *The Cosmic Code*. New York: Simon and Schuster.

Pearle, Philip (1970) Hidden-variable example based upon data rejection. *Physical Review D* 2, 1418–25.

Peirce, Charles Sanders (1933) The logic of quantity. In *Collected Papers*, vol. 4, ed. C. Hartshorne and P. Weiss. Cambridge, MA: Harvard University Press, pp. 59–131.

Pilley, John (1933) *Electricity*. Oxford: Clarendon Press.

Place, U. T. (1996) contributions in Armstrong et al. (1996).

Poynting, John Henry (1920) On the transfer of energy in the electromagnetic field. In *Scientific Papers*. Cambridge, UK: Cambridge University Press, pp. 175–93.

Poynting, John Henry, and Thomson, J. J. (1928) *A Text-book of Physics*, vol. 3, 9th edn. London: Griffen.

Preston, S. Tolver (1885) On some electromagnetic experiments of Faraday and Plucker. *Philosophical Magazine* 19, 131–40.

Price, H. H. (1953) *Thinking and Experience*. Cambridge, MA: Harvard University Press.

Price, Huw (1996) *Time's Arrow and Archimedes' Point*. New York: Oxford University Press.

Prior, A. N. (1953) Three-valued logic and future contingents. *Philosophical Quarterly* 3, 317–26.

Pugh, Emerson M., and Pugh, George E. (1967) Physical significance of the Poynting vector in static fields. *American Journal of Physics* 35, 153–6.

———— (1970) *Principles of Electricity and Magnetism*, 2nd edn. Reading, MA: Addison-Wesley.

Purcell, Edward M. (1965) *Electricity and Magnetism*. New York: McGraw-Hill.

Putnam, Hilary (1975) The analytic and the synthetic. In *Mind, Language, and Reality: Philosophical Papers*, vol. 2. Cambridge, UK: Cambridge University Press, pp. 33–69.

Redhead, Michael (1975) Symmetry in intertheory relations. *Synthese* 30, 77–112.

———— (1982) Quantum field theory for philosophers. In P. D. Asquith and T. Nickles (eds.), *PSA 1982*, vol. 2, pp. 57–99.

———— (1987) *Incompleteness, Nonlocality, and Realism*. Oxford: Clarendon Press.

Reichenbach, Hans (1949) The philosophical significance of the theory of relativity. In P. A. Schilpp (ed.), *Albert Einstein: Philosopher-Scientist*. New York: Tudor, pp. 289–311.

———— (1958) *The Philosophy of Space and Time*. New York: Dover.

Resnick, Robert, and Halliday, David (1985) *Basic Concepts in Relativity and Early Quantum Theory*, 2nd edn. New York: Wiley.

Robinson, F. N. H. (1994) Comment on a recent paper by Clark Jeffries. *SIAM Review* 36, 633–7.

———— (1995) *An Introduction to Special Relativity and its Applications*. Singapore: World Scientific.

Roche, John (1987) Explaining electromagnetic induction: a critical re-examination. *Physics Education* 22, 91–9.

Rohrlich, Fritz (1965) *Classical Charged Particles*. Reading, MA: Addison-Wesley.

———— (1987) *From Paradox to Reality*. Cambridge, UK: Cambridge University Press.

Romer, Robert (1966) Angular momentum of static electromagnetic fields. *American Journal of Physics* 34, 772–8.

————— (1982) Alternatives to the Poynting vector for describing the flow of electromagnetic energy. *American Journal of Physics* 50, 1166–8.

————— (1995) Electromagnetic field momentum. *American Journal of Physics* 63, 777–9.

Roseveare, N. T. (1982) *Mercury's Perihelion from Le Verrier to Einstein*. Oxford: Clarendon Press.

Rosser, W. Geraint V. (1997) *Interpretation of Classical Electromagnetism*. Dordrecht: Kluwer.

Russell, Bertrand (1927) *The Analysis of Matter*. New York: Harcourt.

————— (1948) *Human Knowledge: Its Scope and Limits*. New York: Simon and Schuster.

Ryle, Gilbert (1949) *The Concept of Mind*. London: Hutchinson.

Salmon, Wesley (1980) *Space, Time, and Motion*. Minneapolis: University of Minnesota Press.

Schurz, Gerhard (1999) Explanation as unification. *Synthese* 120, 95–114.

Sciama, D. W. (1983) The role of particle physics in cosmology and galactic astronomy. In G. O. Abell and G. Chincarini (eds.), *Early Evolution of the Universe and its Present Structure*. Dordrecht: Reidel, pp. 493–506.

Shimony, Abner (1988) The reality of the quantum world. *Scientific American* 258 (1), 46–53. (European edition, 36–43.)

Shoemaker, Sidney (1969) Time without change. *Journal of Philosophy* 66, 363–81.

Siegel, Daniel (1985) Mechanical image and reality in Maxwell's electromagnetic theory. In P. M. Harman (ed.), *Wranglers and Physicists*. Manchester: Manchester University Press, pp. 180–201.

Skilling, Hugh (1948) *Fundamentals of Electric Waves*, 2nd edn. New York: Wiley.

Sklar, Lawrence (1976) *Space, Time, and Space-Time*. Berkeley: University of California Press.

Smart, J. J. C. (1963) *Philosophy and Scientific Realism*. London: Routledge.

Sober, Elliott (1982) Dispositions and subjunctive conditionals, or, Dormitive virtues are no laughing matter. *Philosophical Review* 91, 591–6.

Sosa, Ernest, and Tooley, Michael (1993) *Causation*. New York: Oxford University Press.

Spector, Marshall (1972) *Methodological Foundations of Relativistic Mechanics*. Notre Dame, IN: University of Notre Dame Press.

Stachel, John (1982) Einstein, Michelson, the context of discovery, and the context of justification. *Astronomische Nachrichten* 303, 47–53.

————— (1987) Einstein on the theory of relativity. In J. Stachel (ed.), *The Collected Papers of Albert Einstein*, vol. 2. Princeton, NJ: Princeton University Press, pp. 253–74.

Taylor, Edwin, and Wheeler, John Archibald (1966) *Spacetime Physics*. San Francisco: Freeman.

Teller, Paul (1989) Relativity, relational holism, and the Bell inequalities. In J. Cushing and E. McMullin (eds.), *Philosophical Consequences of Quantum Theory*. Notre Dame, IN: University of Notre Dame Press, pp. 208–23.

―――― (1995) *An Interpretive Introduction to Quantum Field Theory*. Princeton, NJ: Princeton University Press.

Telushkin, Joseph (1992) *Jewish Humor*. New York: Morrow.

Thomson, J. J. (1886) Report on electrical theories. In *BAAS Report – 1885*. London: John Murray, pp. 97–155.

―――― (1925) *The Structure of Light*. Cambridge, UK: Cambridge University Press.

Thomson, William [Lord Kelvin] (1872) *Reprint of Papers on Electrostatics and Magnetism*. London: Macmillan.

―――― (1987) *Kelvin's Baltimore Lectures and Modern Theoretical Physics*, ed. R. Kargon and P. Achinstein. Cambridge, MA: MIT Press.

Thorne, Kip (1994) *Black Holes and Time Warps*. New York: W.W. Norton.

Tipler, Paul (1976) *Physics*. New York: Worth.

Tittel, Wolfgang, Brendel, J., Gisin, Nicolus, and Zbinden, H. (1999) Long-distance Bell-type tests using energy–time entangled photons. *Physical Review A* 59, 4150–63.

Van Fraassen, Bas (1980) *The Scientific Image*. Oxford: Clarendon Press.

―――― (1991) *Quantum Mechanics: An Empiricist View*. Oxford: Clarendon Press.

Viviani, A., and Viviani R. (1977) Comment on "Spinning magnetic fields." *Journal of Applied Physics* 48, 3981–3.

Von Laue, Max (1949) Inertia and energy. In P. A. Schilpp (ed.), *Albert Einstein: Philosopher-Scientist*. New York: Tudor, pp. 501–34.

Weihs, Gregor, Jennewein, Thomas, Simon, Christoph, Weinfurter, Harald, and Zeilinger, Anton (1998) Violations of Bell's inequalities under strict Einstein locality conditions. *Physical Review Letters* 81, 5039–43.

Weinberg, Steven (1993) *Dreams of a Final Theory*. New York: Pantheon.

Weyl, Hermann (1922) *Space, Time, Matter*. New York: Dover.

Wheeler, John Archibald, and Zurek, Wojciech Hubert (eds.) (1983) *Quantum Theory and Measurement*. Princeton, NJ: Princeton University Press.

Whewell, William (1967) *The Philosophy of the Inductive Sciences*, 2nd edn. New York: Johnson Reprint.

Wigner, Eugene (1967) *Symmetries and Reflections*. Bloomington: Indiana University Press.

Wise, M. Norton (1981) The flow analogy to electricity and magnetism Part 1: William Thomson's reformulation of action at a distance. *Archive for History of Exact Sciences* 25, 19–70.

Zahar, Elie (1989) *Einstein's Revolution: A Study in Heuristic*. La Salle, IL: Open Court.

Zink, J. W. (1967) Potential energy and the hydrogen atom. *American Journal of Physics* 35, 771–3.

Index